T0172243

SpringerBriefs in Applied Sciences and Technology

Thermal Engineering and Applied Science

Series Editor

Francis A. Kulacki, Department of Mechanical Engineering, University of Minnesota, Minneapolis, MN, USA

More information about this series at http://www.springer.com/series/10305

Sujoy Kumar Saha • Hrishiraj Ranjan
Madhu Sruthi Emani • Anand Kumar Bharti

Two-Phase Heat Transfer Enhancement

 Springer

Sujoy Kumar Saha
Mechanical Engineering Department
Indian Institute of Engineering
Science and Technology, Shibpur
Howrah, West Bengal, India

Hrishiraj Ranjan
Mechanical Engineering Department
Indian Institute of Engineering
Science and Technology, Shibpur
Howrah, West Bengal, India

Madhu Sruthi Emani
Mechanical Engineering Department
Indian Institute of Engineering
Science and Technology, Shibpur
Howrah, West Bengal, India

Anand Kumar Bharti
Mechanical Engineering Department
Indian Institute of Engineering
Science and Technology, Shibpur
Howrah, West Bengal, India

ISSN 2191-530X   ISSN 2191-5318   (electronic)
SpringerBriefs in Applied Sciences and Technology
ISSN 2193-2530   ISSN 2193-2549   (electronic)
SpringerBriefs in Thermal Engineering and Applied Science
ISBN 978-3-030-20754-0   ISBN 978-3-030-20755-7   (eBook)
https://doi.org/10.1007/978-3-030-20755-7

This Springer imprint is published by the registered company Springer Nature Switzerland AG.
The registered company address is: Gewerbestrasse 11, 6330 Cham, Switzerland

Contents

Nomenclature

A Area
A_{ni} Nominal surface area based on fin root diameter
D Diameter
d Diameter
d_i Fin root diameter
d_o Outer diameter
E Electric field strength
e Fin height
f Friction factor
f_e EHD force density
G Mass velocity
g Gravitational acceleration
h Convective heat transfer coefficient
h_{lv} Latent heat
K Thermal conductivity
L Length
L Characteristics dimension, length
M_d Masuda number
N Number of channels
n_s Number of fins
Nu Nusselt number
P Pressure
ΔP Pressure drop
p Pressure
p_f Fin pitch normal to the fins
Q Heat transfer rate
q Heat flux
Re Reynolds number
Re_1 Liquid Reynolds number
T Temperature

t	Time
U	Mean velocity
V	Volume
X	Effective quality
x	Vapour quality

Greek Symbols

α	Heat transfer coefficient, apex angle of the fin
β	Helix angle
δ	Liquid film thickness
Δp	Pressure drop
ΔT	Temperature difference
ε	Permittivity
μ	Dynamic viscosity
ρ	Density
σ	Surface tension

Subscripts

ave	Average
ev	Evaporation
in	Inlet
l	Liquid
s	Saturated
sub	Subcooled
v	Vapour

Abbreviation

CHF	Critical heat flux
CNT	Carbon nanotube
EHD	Electrohydrodynamic
ONB	Onset of nucleate boiling

Chapter 1
Introduction

1.1 Development of an Enhanced Boiling Surface and the Pioneering Studies of the Fundamentals

High-performance nucleate boiling is aided by special surface geometries enhancing heat transfer. Heat transfer coefficients are substantially increased by special surface geometries. However, the boiling coefficient is defined in terms of projected or flat plate surface area. Earlier, the special surfaces having roughness-enhanced heat transfer lasted for few hours. But, due to ageing, that enhancement used to be withered away. This caused vigorous research to understand the character of boiling sites and the geometry of surface necessary for stable vapour traps. Thereafter, the preparation of nucleate boiling surface geometries progressed over the years. Subsequently, as a result of vigorous research, correlations for the prediction of the boiling coefficient have been developed as well as understanding boiling mechanism has been facilitated. An explosion of journal and conference papers has occurred, and useful patents have been granted. This clearly bears the testimony to the improvement of enhanced boiling surfaces. Academic research has also been instrumental to the understanding of the physical mechanism of the enhanced surfaces.

Jakob (1949) and Sauer (1935) tell us that surface finish obtained by sandblasting method and a square grid of machined groves is responsible for evanescent improvement of nucleate boiling heat transfer coefficient. However, this short-living performance improvement quickly decays due to an ageing effect. The cavities formed by the roughening of surfaces are not able to provide stable vapour traps required for sustained nucleation process. Berenson (1962), Kurihara and Myers (1960) and Corty and Foust (1955) tried to boil several liquids like water and several organic liquids on flat surfaces roughened with emery paper of different coarseness and lapping compound. They have observed increased boiling coefficient from an increased area density of nucleation sites. However, this is also accompanied by surface ageing after short-term improvement. These artificially roughened sites cause boiling incipience at a driving potential obtained due to temperature difference. This observation was the motivation behind making the artificial sites responsible for early incipience of boiling at low ΔT and stable vapour traps.

Research in the above line and direction was simultaneously accompanied by parallel fundamental studies of the character of nucleation sites, the effect of cavity shape and theoretical models of the nucleation process. Naturally occurring pits and scratches on boiling surfaces are active boiling sites. High-speed photographic studies enabled by modern technology bears the evidence. Clark et al. (1959) and Griffith and Wallis (1960) have explained that the cavity geometry is important in two ways. The diameter of the mouth of the cavity determines the superheat needed to initiate boiling, and its shape determines its stability once boiling has begun. The signature of liquid-vapour interface in a re-entrant cavity and the contact angle are responsible for maintaining the vapour nucleus in the presence of subcooled liquid. It is, therefore, understood that a re-entrant cavity must be a very stable vapour trap.

Benjamin and Westwater (1961) studied the performance of re-entrant cavity as a vapour trap. They observed that the boiling site remained active after surface subcooling, whereas the naturally occurring sites were flooded. Yatabe and Westwater (1966) observed that interior shape of a re-entrant cavity did not have strong influence on the boiling performance of the site. Hsu (1962), Bankoff (1958) and Moore and Mesler (1971) developed some working models for incipient boiling on a heated surface, and they concluded that high vaporization rates occurred across a microlayer of liquid at the base of the bubble. Even though these above-mentioned investigations lay the foundation of the main prop of the underlined physics, the later research has further advanced the technology of nucleation of discrete cavities. Nevertheless, the state-of-the-art enhanced surfaces generally consist of 2D re-entrant grooves or interconnected cavities. Several researchers all over the world have made significant contributions on the study of the effect of re-entrant cavity shape.

References

Bankoff SG (1958) Entrapment of gas in the spreading of a liquid over a rough surface. AICHE J 4(1):24–26

Benjamin JE, Westwater JW (1961) Bubble growth in nucleate boiling of a binary mixture. ASME International Developments in Heat Transfer, New York, pp 212–218

Berenson PJ (1962) Experiments on pool-boiling heat transfer. Int J Heat Mass Transf 5(10):985–999

Clark H, Strenge PS, Westwater JW (1959) Active sites for nucleate boiling. Chem Eng Prog 1:55

Corty C, Foust AS (1955) Surface variables in nucleate boiling. Chem Eng Prog Symp Ser 51(17):1–12

Griffith P, Wallis JD (1960) The role of surface conditions in nucleate boiling. Chem Eng Frog Symp Ser 56(49):49–63

Hsu YY (1962) On the size range of active nucleation cavities on a heating surface. J Heat Transf 84(3):207–213

Jakob M (1949) Heat transfer. Wiley, New York, pp 636–638

Kurihara HM, Myers JE (1960) The effects of superheat and surface roughness on boiling coefficients. AICHE J 6(1):83–91

Moore FD, Mesler RB (1971) The measurement of rapid surface temperature fluctuations during nucleate boiling of water. AICHE J 7(4):620–624

Sauer ET (1935) MS thesis, Department of Mechanical Engineering, Massachusetts institute of technology, Cambridge, MA

Yatabe JM, Westwater JW (1966) Bubble growth rates for ethanol-water and ethanol-isopropanol mixtures. Chem Eng Prog Symp 62-64:17–23

Chapter 2
Pool Boiling Enhancement Techniques

2.1 Abrasives, Open Groves, Three-Dimensional Cavities

Several surveys of enhancement techniques dealing with the development of enhanced boiling surfaces have been made by Webb (1981, 1983), Webb and Donald (2004), Dundin et al. (1990) and Matijević et al. (1992).

Abrasive treated surfaces were examined by Chaudhri and McDougall (1969). Emery roughening and sand blasting for boiling were tested by Luke (1997). Kang (2000) investigated the effect of sand paper roughness on boiling. Long-term ageing effects were observed by all the investigators. Also, the enhancement depends on tube orientation (horizontal or vertical) and L/D ratio of the tube. Increased liquid agitation with bubble movement has pronounced effect on boiling. Scratches by dragging a sharp-pointed scriber across a polished surface from parallel groove and the spacing of the scratches influence the boiling performance. There is an optimum scratch spacing for the performance (Bonilla et al. 1965).

Three-dimensional cavities were investigated by Griffith and Wallis (1960). They have observed that fundamental characteristics like boiling incipience and stability greatly depend on the diameter of the conical cavity. Addition of a non-wetting coating within the cavity causes incipience of boiling at a much lower superheat. Non-wetting phenomenon is beneficial for high surface tension fluids like water, etc. Marto et al. (1968) and Griffith and Wallis (1960) have observed that the cavity shape has a profound effect on boiling performance. Re-entrant cavity shape is definitely preferred over a conical or cylindrical shape. This is so because lower superheat is required for incipience, and the re-entrant shape provides a more stable vapour trap. High area density rather than few discrete cavities is required for boiling to occur in nucleation sites.

2.2   Electroplating, Attached Wire and Screen Promoters, Pierced Three-Dimensional Cover Sheets, Etched Surfaces

Electroplating was studied by Bliss Jr et al. (1969) and Albertson (1977). The heat transfer either improves or deteriorates depending on which metal coatings are given. They have observed that the thermal properties have no conceivable direct impact on boiling performance; rather the better or worse performance is attributed to the formation of dendrites or nodules on the base surface. High current densities during electroplating and cold rolling of the tube partially compact the nodules, and this improves the boiling performance.

Corman and McLaughlin (1976) experimented with felt-metal nickel and copper wicking of various thicknesses. Heat flux influences the boiling coefficient inversely. Increasing the heat flux, however, takes a crossover point; the wick surface becomes inferior, but after that, the slope of the boiling curve, though remaining lower, gives higher critical flux than the smooth surface. Thinner wick structures give a better thermal performance. Hasegawa et al. (1975) observed that larger mesh openings give more improvement in critical heat flux (CHF), and in general, the screen increases CHF; also one screen layer is better than two layers. Asakavicius et al. (1979), Danilova and Tikhonov (1996), Tsay et al. (1996), Liu et al. (2001) and Guglielmini et al. (1988) have worked on the pool boiling performance of enhanced surface, and they have observed that greater mesh screens "block the bubbles on the surface" and the enhancement openings in the screen close to the bubble departure diameter give better boiling performance. Figure 2.1 shows the loosely fitted metallic (or non-metallic) wire-wrapped integral fin tube. Webb (1970), Uma et al. (2000), Palm (1992), Ragi (1972) and Kim (1996) worked on the improved pool boiling performance using pierced three-dimensional cover sheets. Figure 2.2 shows water boiling on surface having re-entrant cavities. Etched surfaces formed by applying chemicals have been studied by Vachon et al. (1969). Laser beam was used by Chu and Moran (1977) to form re-entrant shaped cylindrical cavities (Fig. 2.3). Chien and Hwang (2012) proposed a novel enhanced surface for boiling heat transfer

Fig. 2.1 Nucleation sites formed by metallic (or non-metallic) wire wrap in-between the fins of an integral tube (Webb and Kim (2005)

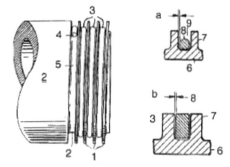

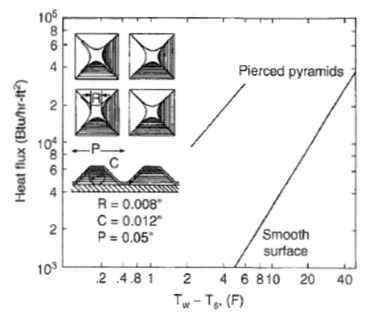

Fig. 2.2 Water boiling at 101 kPa (1 atm) on surface having re-entrant cavities formed by brazing pierced foil sheet to base surface (from Ragi 1972)

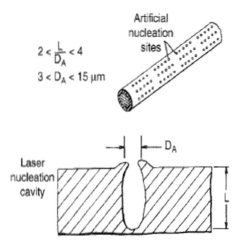

Fig. 2.3 Re-entrant cavities formed by laser beam (from Chu and Moran 1977)

augmentation. A wire mesh has been wrapped on low fin tubes. They observed that the effect of variation of fin geometry was profound in mesh-attached fin tubes. Komendantov et al. (2004) studied straight and spiral tubes for heat transfer at boiling crisis using R113 as working fluid.

2.3   Coatings, Porous Surfaces, Structured Surfaces, Combined Structured and Porous Surfaces, Composite Surfaces and Pool Boiling Tests of Enhanced Surfaces

A thin non-wetting coating on the interior walls of a nucleating cavity enhances its stability on a boiling nucleating site. A larger liquid–vapour interface radius within the cavity obtained by the coating requires less superheat (Griffith and Wallis 1960). Unwetted cavities are more stable than wetted cavities. Gaertner (1967), Hummel (1965), Young and Hummel (1965), Bergles et al. (1968) and Vachon et al. (1969) applied this coating of non-wetting material on a pool boiling surface. Closely packed nucleation sites were formed by needle of sharp punches and parallel scratches. A coating material of low surface energy is given over the surface containing the artificial sites. This is then removed from the flat face by abrasion; a thin film of the material is deposited in each cavity, and the boiling site remains active for a much longer time. When the bubble is in contact with the non-wetted surface, much lower superheat is required for its existence and initial growth since the bubble has a much larger radius of curvature than that for a wetted surface. For bubble growth in diameter past the area of unwetted spot, it quickly reverts to the spherical shape and grows very rapidly due to evaporation of the thin liquid film at the base of the spherical bubble. A Teflon spotting method is effective for such surface–liquid combination that gives larger contact angles. For coating thicknesses larger than an optimum value, heat transfer decreases due to the insulating effect of the thick coating. Few micron coating layer of oxides and ceramics as observed in scanning electron microscope gives better boiling performance (Zhou and Bier 1997; Sridharan et al. 2002; Uhle 1998; Jamialahmadi and Müller-Steinhagen 1993; Imadojemu et al. 1995). Dizon et al. (2004) proposed metallic microporous coating for enhancing critical heat flux for pool boiling. Wasekar and Manglik (2017) have observed that use of additives made of polymers at low concentrations augmented the nucleate pool boing heat transfer. This behaviour has been resulted due to change in surface tension and viscosity of the solvent on addition of surfactants and additives. Vasiliev et al. (2012) carried out experimental study on pool boiling and evaporative heat transfer in mini- and microchannels which have porous coatings and found considerable enhancement in heat transfer rates.

The coating of a fraction of a millimetre thickness of a sintered porous metallic matrix bonded to the base surface improves boiling performance and increases CHF (Milton 1968, 1970, 1971; Gottzmann et al. 1971, 1973; Kim and Bergles 1988). The sintered particle coating is made by mixing the particles into liquid slurry containing a temporary binder. The surface is then coated with the slurry which is air-dried and then sintered in a baking oven. The sintering is done at a temperature slightly below the particle melting point. However, sintering is detrimental to boiling performance if it anneals the base surface. Figure 2.4 and Tables 2.1 and 2.2 show the details of porous coating tested. Further detailed studies may be obtained from Nishikawa (1983), Nishikawa and Ito (1980), O'Neill et al. (1972), Czikk and O'Neill (1979), Griffith and Wallis (1960), Grant (1977), Hausner and Mal (1982),

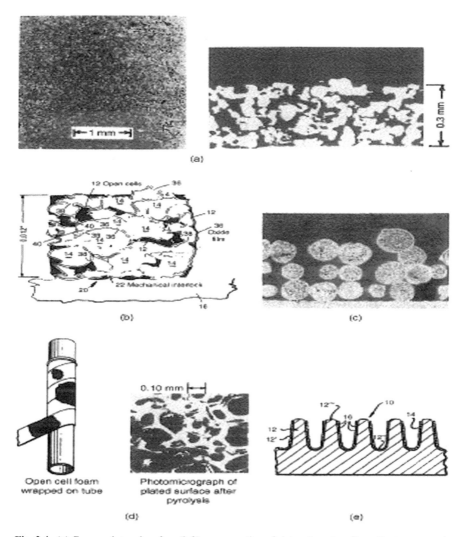

Fig. 2.4 (**a**) Copper-sintered surface (left); cross-section of sintered coating (from Gottzmann et al. 1971, 1973), (**b**) cross-sectional photograph of aluminium flame-sprayed surface (from Dahl and Erb 1976), (**c**) 0.115-mm-diameter spherical particles tested by Fujji et al. (1979), (**d**) porous coating formed by plating copper on polyurethane foam-wrapped tube (from Janowski et al. 1978), (**e**) porous coating on fins formed by electroplating in a bath containing graphite particles (from Shum 1980)

Hsieh and Weng (1997), Hsieh and Yang (2001), Shum (1980), Dahl and Erb (1976), Modahl and Luckeroth (1982), Sanborn et al. (1982), Liu et al. (1987), Janowski et al. (1978), Sachar and Silvestri (1983), Zohler (1990), Chang and You (1997) and O'Connor and You (1995).

   The pore size distribution may be obtained from a quantitative scanning microscope by examining cross-sectional slices of the coating. This statistically defined

Table 2.1 Porous coating tested (Webb and Kim 2005)

References	Particle metal	Test fluid	Coating structure
O'Neill et al. (1972)	Cu, Ni, Al	R-11 and others	Sintered particles (non-uniform size and shape)
Nishikawa (1983)	Cu, bronze	R-11, R-113, benzene	Sintered spherical particles
Fujii et al. (1979)	Cu	R-11, R-113	Spherical particles bonded by electroplating
Dahl and Erb (1976)	Al	R-113	Metal sprayed powder
Janowski et al. (1978)	Cu	No published data	Electroplated polyurethane foam
Kajikawa et al. (1983)	Cu	R-11	Sintered steel fibres
Yilmaz and Westwater (1981)	Cu	Isopropanol, p-xylene	Metal spray with copper wire
Kartsounes (1975)	Cu	R-12	Metal sprayed coating

Table 2.2 Porous copper coating (Webb and Kim 2005)

References	d_p (µm)	δ (mm)	δ/d_p	ε
Kajikawa et al. (1983)	12	1.0	8	0.65
Milton (1968)	<44	0.25	6	
Fujii et al. (1979)	115	0.40	3–4	0.49
Nishikawa (1983)	250	2.0	8	0.66
Milton (1968)	250–500	1.1	3	
Nishikawa (1983)	500	2.0	4	0.42

pore size distribution may, however, not be wholly realistic because the method does not differentiate between connected and closed pores. A more acceptable method involves the use of a capillary rise test using a wetting fluid near zero contact angles. Essentially all vaporization occurs within the porous matrix and high performance is because (1) the porous structure entraps large radius vapour–liquid interfaces reducing the superheat required for nucleation and (2) the porous structure provides a much larger surface area for thin film or microlayer evaporation than that exists with a flat surface. Structured surfaces are integral roughness, and they employ cold metal and form a high area density of re-entrant nucleation sites interconnected below the surface. These are re-entrant grooves or tunnels, and these are not discrete cavities (Fig. 2.5). Pais and Webb (1991), Webb (1972), Saier et al. (1979), Fujie et al. (1977), Arai (1977), Torii et al. (1978), Bell and Mueller (1984), Thors et al. (1997), Kim and Choi (2001), Chien and Webb (1998a, b, c, d, e), Chien and Chang (2004), Chien and Chen (2000) and Kun and Czikk (1969) deal with the details of boiling performance by using structured surfaces. Figures 2.6, 2.7, and 2.8 give the views of structured surface. Arias and Reventos (2010) studied the impact of lift forces on boiling based on the Taylor–Helmholtz instabilities for the liquid–vapour interface.

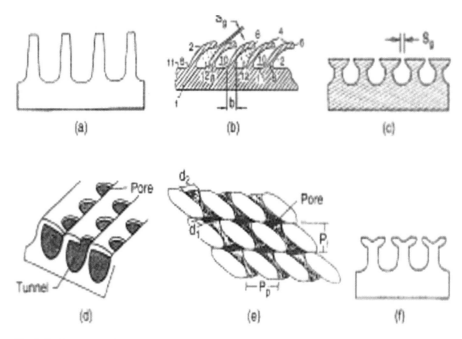

Fig. 2.5 Six commercially available enhanced boiling surfaces: (**a**) integral-fin tube, (**b**) Trane bent fin, (**c**) Wieland GEWA-TW, (**d**) Hitachi Thermoexcel-E, (**e**) Wolverine Turbo-B II and (**f**) Wieland GEWA-SE (from Webb and Kim 2005)

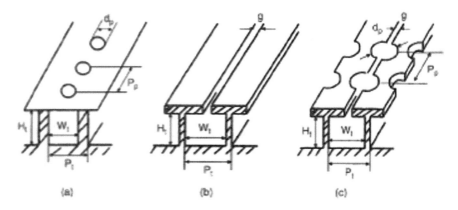

Fig. 2.6 Possible basic surface/substance structures for enhanced boiling tubes made from an integral-fin tube: (**a**) surface pores, (**b**) surface gaps, (**c**) surface pores with connecting gaps (from Kim and Choi 2001)

Ma et al. (1986) have worked with the combination of structured and porous surfaces, and they have found quite high performance. Cross-sectional shape may be parallel rectangular grooves covered by fine mesh screens, mesh screen covered by a thin metal plate having pores, whereas the cross-sectional shape of the grooves may

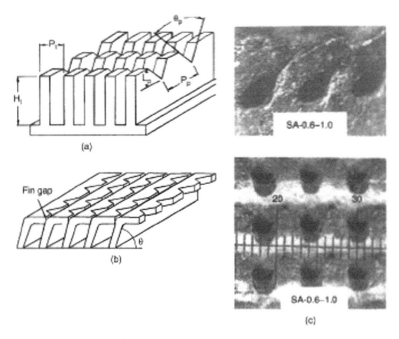

Fig. 2.7 Structured surfaces made by Chien and Chang (2003). (**a**) Unbent fins, (**b**) bent fins, (**c**) cross section through fins of surface SA 0.6–1.0 (0.6-mm fin pitch and pore pitch with 1.0-mm-high fins)

even be V shaped. The screen porosity may vary. Antonelli and O'Neill (1981) have observed that the surface performance is competitive with the high heat flux surface. Triangular grooves provide higher heat flux than rectangular grooves at high heat flux. The high performance may, however, be accompanied with high expenses also. Moderate heat transfer enhancement may be obtained by using a composite fibre–Cu surface (Yang et al. 1996). The reason for heat transfer enhancement is the hotspots on the graphite tips which act as nucleation activators. A composite surface formed by embedding graphite fibres in Cu or Al substrates has been used by Liang and Yang (1998). High graphite thermal conductivity gives higher temperature at graphite–liquid surface, for which the very small microcavities in the graphite become active. Chien and Chang (2004) studied surfaces with bending fins that have triangular cuts for pooling boiling heat transfer enhancement. The geometry of the surface has lesser effect heat transfer for HFC-4310 as compared to that for water as the surface tension and latent heat of HFC-4310 is less than that of water. Krikkis et al. (2003) analysed the role of pin-fin in multiboiling heat transfer mode. The regions of convective heat transfer, nucleate boiling, transition and film boiling heat transfer have been considered. Orman (2016) used pin-fin microstructures to study the pool boiling heat characteristics with distilled water and ethyl alcohol as test fluids. They observed that microstructures with smaller inter-fin space and larger heights have shown best performance. The enhancement has been observed to be

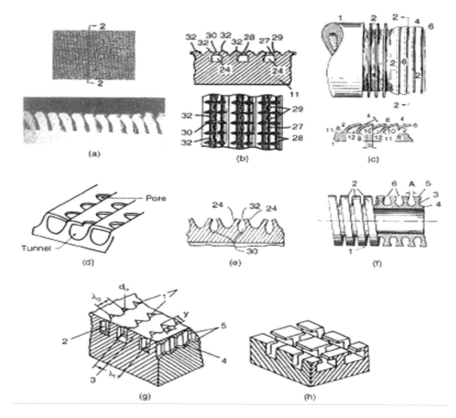

Fig. 2.8 Structured boiling surfaces: (**a**) cross-grooved surface having re-entrant cavities (Kun and Czikk 1969), (**b**) deeply knurled Y-shaped fins (Szumigala 1971), (**c**) re-entrant grooves formed by bending tips of integral fin tube (Webb 1972), (**d**) re-entrant grooves having minute spaced holes at the top of tunnel (Fujie et al. 1977), (**e**) re-entrant grooves formed by bending of fins of unequal heights (Brothers and Kalifelz 1979), (**f**) re-entrant grooves formed by flattening fin tips of integral fin tube (Saier et al. 1979), (**g**) JK-2 tubes tested by Zhang et al. (1992), (**h**) geometry developed by Zhang and Dong (1992)

almost eight times over the performance of smooth tube. Yilmaz and Westwater (1981), Li et al. (1992), Thors et al. (1997), Tatara and Payvar (2000), Webb and Pais (1992), Wang et al. (1998), Saidi et al. (1999), Kim and Choi (2001), Chien and Chang (2004), Kedzierski (1995), Srinivasan et al. (2001), Tarrad and Burnside (1993), Sokol et al. (1990), Hübner and Künstler (1997), Mertz et al. (2002) and Kulenovic et al. (2002) have done several single-tube pool boiling tests of enhanced surfaces. Figure 2.9 shows comparative single-tube pool boiling results for isopropyl alcohol at 101 kPa (Yilmaz and Westwater 1981). Figure 2.10 shows the progression of boiling tube performance from old to new versions of the Turbo-B-type tube (Thors et al. 1997). Figure 2.11 shows the photograph of 1654 fpm tube made by Kim and Choi (2001) and commercial turbo-tube.

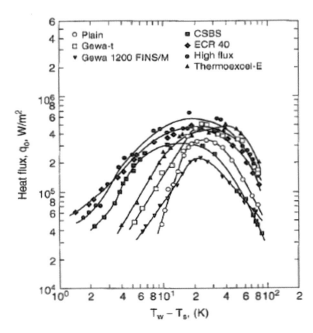

Fig. 2.9 Comparative single-tube pool boiling test results for isopropyl alcohol at 101 kPa (1 atm) (from Yilmaz and Westwater 1981)

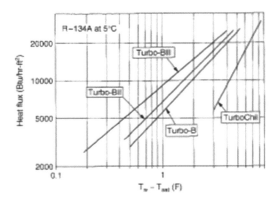

Fig. 2.10 Progression of boiling tube performance (R-134a at 5.0 °C) from old to new versions of the Turbo-B-type tube, Tube I (TurboChil), Tube II (Turbo-B), Tube III (Turbo-BII), Tube IV (Turbo-BIII) (Thors et al. 1997) Progression of boiling tube performance (R-134a at 5.0 °C) from old to new versions of the Turbo-B-type tube Tube I (TurboChil), Tube II (Turbo-B), Tube III (Turbo-BII), Tube IV (Turbo-BIII) (from Thors et al. 1997)

Fig. 2.11 (a) Photograph of 1654 fins/m tube made by Kim and Choi (2001) having 0.2-mm-diameter surface pores and 0.04-mm-gaps. (b) Photograph of commercial Turbo-BIII tube

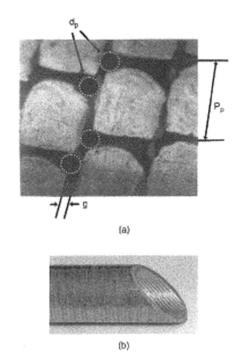

(a)

(b)

2.4 Fundamental Theory, Effects of Boiling Hysteresis and Orientation, Boiling Mechanism and Models for Structured Surfaces, Porous Surfaces, Critical Heat Flux and Thin Film Evaporation

Pool boiling on enhanced surfaces can be understood only when liquid superheat, effect of cavity shape and contact angle on superheat, entrapment of vapour in cavities, effect of dissolved gases, nucleation at a surface cavity, bubble departure diameter and bubble dynamics are clearly known. Even though vigorous research has been done for a sufficient length of time and several research papers have been published on this topic, the complete physics of pool boiling on enhanced surfaces remains to be understood thoroughly. To name a few, the following publications may be looked into which will shed some light on the theory and physics of pool boiling on enhanced surfaces: Carey (1992), Rohsenow (1985), Webb (1983), Lorenz et al. (1974), Hsu (1962), Fritz (1935), Chien and Webb (1998c), Kulenovic et al. (2002), Kurihara and Myers (1960), Mikic and Rohsenow (1969), Nakayama et al. (1980a), Haider and Webb (1997), Bergles and Chyu (1982), Kim and Bergles (1988), Ko et al. (1992), Liu et al. (1987), Bar-Cohen (1992), Malyshenko and Styrikovich (1992), Rainey and You (2001) and Chang and You (1997).

The internal shape of the cavity has a significant influence on the liquid superheat required for liquid–vapour interface to exist within a cavity (Fig. 2.12). The radius of

Fig. 2.12 Variation of
bubble radius during growth
for: (**a**) $8 < cjJ < 90°$,
(**b**) $2cjJ < 8 < 90°$,
(**c**) $8 > cjJ + 90°$,
(**d**) variation with cavity
volume (from Carey 1992)

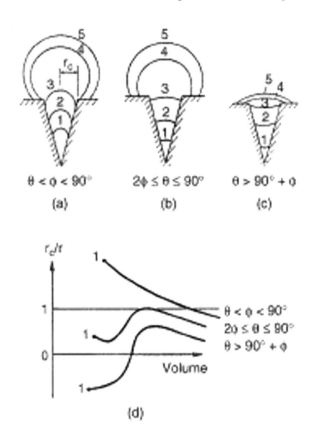

the vapour interface changes with its movement within cavities of various shapes.
Boiling occurs usually with subcooled liquid. The subcooled liquid is heated to the
boiling state when heat is applied. Detailed analysis has not been done yet to explain
how inert gas can be trapped in cavities, when flooded by subcooled liquid. Inert gas
trapped in the cavity influences the boiling process. It may so happen that the vapour
trapped in the cavity may not contain an inert gas. It is perfectly possible that the
charged liquid may contain a dissolved gas coming out of the solution when the
liquid is raised to boiling temperature. This evolved gas may add to the trapped
vapour volume. A concave vapour interface requires less liquid superheat when inert
gas is present in the cavity. When boiling occurs from cavities containing inert gases,
each departing bubble removes a small amount of inert gas from the cavity. As the
time elapses, a higher temperature difference is necessary to sustain boiling since the
partial pressure of the inert gas reduces.

Hsu (1962) proposed a better analytical model to define the wall superheat
necessary to support periodic bubble growth at a plain heated surface. The model
of Hsu (1962) assumes transient heat flux to the cold liquid layer that floods the
surface after bubble departure. When the cavity size is known, the temperature
difference needed to support nucleation from the cavity at a specified heat flux can

easily be obtained. However, the Hsu (1962) model for plain surfaces faces a problem because the size of a natural nucleation site is an unknown quantity. For the enhanced surfaces, the bubbles grow from pores in the surface that are typically much larger than the microcavities associated with plain surfaces.

Bubbles grow periodically and are released from surface. The bubble departs when buoyancy exceeds the surface tension force holding the bubbles to the surface. Bubble diameter can be calculated by a force balance between buoyancy and surface tension forces acting on a departing bubble and when inertia forces are neglected. Several theoretical treatments have been accorded to calculate the bubble departure diameter. Some treatments do account for inertia term in the momentum equation for growing bubble. Figure 2.13 shows vapour–liquid interface radius in cavities of different shapes. Figure 2.14 demonstrates how a cavity may trap air when the surface is flooded with liquid. The beginning of bubble growth and the steady state slug flow patterns in a closed thermosyphon loop has been studied by Wei et al. (2014).

A detailed study of CHF enhancement using both pure fluids and nanofluids has been made by Bi et al. (2015). They have given a composite model for CHF and average heat flux prediction incorporating microlayer evaporation, transient conduction due to bubble departure and micro-convection due to bubble growth and bubble movement. Bubble influence area interface has also been taken into account in the model. The model is capable enough to predict the average heat flux and the critical heat flux (CHF) for surface dryout fraction exceeding a certain value. They calculated nanofluid properties and used contact angles obtained from the experiments. They observed many-fold increase of CHF with the decrease of contact angle from 80° to 20°. Shahmoradi et al. (2013), Jung et al. (2013), Vazquez and Kumar (2013), Jung et al. (2012), Hegde et al. (2012), Kole and Dey (2012), Bolukbasi and Ciloglu (2011) and Gerardi et al. (2011) document substantial enhancement of CHF within nanofluid boiling. However, contradictory results with respect to average wall heat transfer rates have been reported in those references. In the above investigations, it has been observed that nanoparticles much smaller than the surface roughness elements deposit on the relatively uneven surface during boiling. A smoother surface causing degradation of the boiling characteristics is created. The surface smoothening overshadows the thermal conductivity enhancement of the nanofluid. Larger concentration of nanofluids results in a moderate increase in the CHF. The deterioration of heat flux may primarily be attributed to the deposition of nanoparticles on the surface which changes the surface wettability. CHF is enhanced many times due to the build-up of a porous layer of nanoparticles on the surface. Table 2.3 shows the summary of the experimental results of nanofluid boiling as obtained by Wen et al. (2011), Suriyawong and Wongwises (2010), Kwark et al. (2010), Liu et al. (2007) and You et al. (2003). Table 2.3 gives in brief the synopsis of pertinent nanofluid boiling experiments. Figure 2.15 compares the calculated average heat flux obtained by Bi et al. (2015) using their analytical model with the experimental results of pure water and nanofluid boiling of Shahmoradi et al. (2013) and Bang and Chang (2005). The heat transfer coefficient and the Nusselt number at different wall heat fluxes as

Fig. 2.13 Vapour–liquid
interface radius in cavities of
different shapes for
$1\sim = 0.04$ mm and $8 = 15°$
(from Webb 1983)

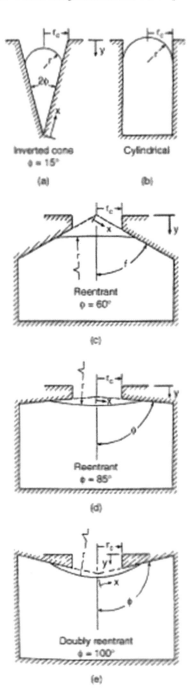

Fig. 2.14 Illustration of how a cavity may trap air when the surface is flooded with liquid. (**a**) Advancing liquid front. (**b**) Liquid–vapour interface for the (**a**) conditions (from Carey 1992)

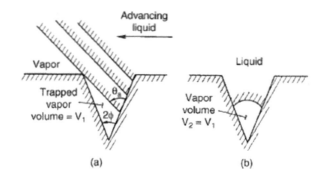

Table 2.3 Synopsis of nanofluid boiling experiments (Bi et al. 2015)

References	Nanofluid properties	Surface property	CHF	Average boiling heat flux
Shahmoradi et al. (2013)	Al_2O_3 (40 nm)-water (0.001–0.1 vol.%)	Flat plate heater (θ from 45° to 20°) (Ra from 5.1 nm to 376 nm; 77 nm to 197 nm)	CHF ↑ 47% for 0.1 vol.%	BHF ↓ 40% for 0.1 vol.%
Jung et al. (2013)	Al_2O_3 (46 nm) + H_2O/LiBr (0.01–0.1 vol.%)	Plate copper heater (θ from 78° ↓ 36°)	CHF ↑ by 48.5% at 0.1 vol.%	BHF ↓ 10%
Vazquez and Kumar (2013)	Silica (10 nm)-water 0.1–2 vol.%	Nichrome wires and ribbons	CHF ↑ 270% at 0.4 vol.%	BHF ↑ 190%
Jung et al. (2012)	Al_2O_3 (45 nm)-water (10^{-5}–0.1 vol.%).	Plate copper heater	CHF ↑ 116%	BHF ↓ 60%
Hegde et al. (2012)	CuO (50 nm)-water, (0.01–0.5%)	NiCr wire (from 0.34 μm to 0.09 μm (0.05 vol.%) and (0.23 μm (0.3%))	CHF ↑ 130% (1.3–3 MW/m²)	BHF ↓ 50%
Kole and Dey (2012)	ZnO (30–40 nm)-ethylene glycol (EG)	Copper cylindrical block (90 nm)	CHF ↑ by 117% at 2.6%	BHF ↑ by 22% at 1.6% and ↓ further loading
Bolukbasi and Ciloglu (2011)	SiO_2 (34 nm)-water (0.001–0.1 vol.%)	Cylindrical brass (typical 90 nm) (Ra ↑ to 620 nm, θ from 70° to 42.07°)	CHF ↑ 28%	BHF ↓ 40%
Gerardi et al. (2011)	Diamond (34 nm) and silica-water (173 nm) (0.1 vol. % for silica and 0.01 vol.% for diamond)	Upward indium-tin-oxide surface (θ from 80° to 16°) (Ra from 30 to 900–2100 nm)	CHF ↑ 100%	BHTC ↓ 50%
Wen et al. (2011)	Al_2O_3 (50–900 nm)-water (0.001–0.1%)	Rectangular brass plates (rough: 420 nm and smooth: 25 nm)	–	Twofold ↑ in BHTC at 0.001% nanofluids

(continued)

Table 2.3 (continued)

References	Nanofluid properties	Surface property	CHF	Average boiling heat flux
Suriyawong and Wongwises (2010)	TiO$_2$-water (21 nm), 0.00005–0.01 vol. %	Horizontal circular plates, copper (*Ra* 0.2 µm) and aluminium (*Ra* 4 µm)	–	BHTC 15% ↑ for copper and a 4% ↑ for aluminium at 0.0001 vol.%; ↓ at higher conc levels
Kwark et al. (2010)	(Al$_2$O$_3$, CuO and diamond)-water (2.7 × 10^{-5}– 2.7 × 10^{-2}% vol.)	1 cm × 1 cm × 0.3 cm copper block	CHF ↑ 31%	BHF no change <500 KW/m^2 and ↓ at high heat flux
Liu et al. (2007)	CuO (30 nm)-water	Copper block	CHF ↑ when conc <1% constant since	BHF ↑ when conc <1%; ↓ >1%
You et al. (2003)	Al$_2$O$_3$-water (0–0.05 g/L)	Polished copper surfaces	CHF ↑ 200%	Same
Das et al. (2003)	Al$_2$O$_3$ (50–150 nm)-water (1–4%)	Rectangular stainless steel (0.4 and 1.15 µm)	–	BHF ↓

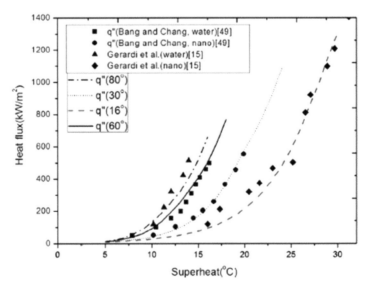

Fig. 2.15 Comparison of calculated average heat flux (Bi et al. 2015) with the experimental results of Shahmoradi et al. (2013) and Bang and Chang (2005)

observed by Bi et al. (2015) have been plotted and compared in Fig. 2.16 with the experimental results of Suriyawong and Wongwises (2010).

The hysteresis effect has been studied by a number of scientists. Hysteresis occurs when heating is initially applied and data are taken in the order of increasing heat

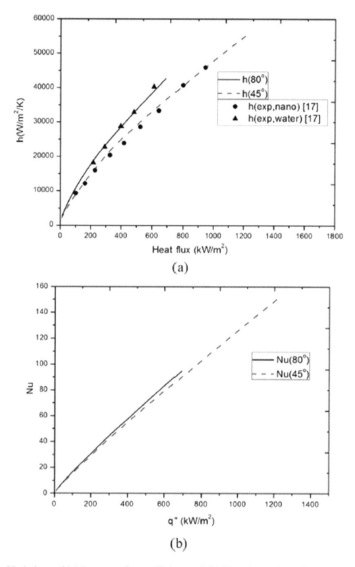

Fig. 2.16 Variations of (**a**) heat transfer coefficient and (**b**) Nusselt number with average wall heat flux predicted by Bi et al. (2015) for both pure water and nanofluid boiling experiments of Suriyawong and Wongwises (2010)

flux. Boiling can occur when a certain amount of superheat is applied after the heat flux starting from zero. Following this, the surfaces suddenly explode into nucleate boiling, and at this point, the surface temperature decreases. Bergles and Chyu (1982) have measured hysteresis effect on the high flux surface for boiling of R113 having single electrically heated tube. The possibility of cavity flooding is more likely for low surface tension wetting fluids than that for a less wetting fluid such as water. The cavity geometry and the existence of dissolved gas have a significant effect on hysteresis. Kim and Bergles (1988) extended their evaluation of hysteresis effects on porous boiling surfaces. Ko et al. (1992) measured hysteresis effect for water and R113 on four different porous coatings having different particle sizes. Porous structures having small pores have the largest incipient superheat. Liu et al. (1987) observed decreasing hysteresis effect as the number of screen layers increase from one to five. Ma et al. (1986) observed that combination of structured and porous surfaces have substantially less hysteresis with methanol compared to R113 as observed by Bergles and Chyu (1982). Bar-Cohen (1992) and Malyshenko and Styrikovich (1992) observed a peculiar hysteresis phenomenon for boiling on porous coatings made of low thermal conductivity particles. Formation of a vapour or gas nucleus in potentially vapour-trapping cavities is due to initial gas trapping, dissolution of dissolved gas, blowing vapour or inert gas into the cavity from an external surface and heterogeneous nucleation. Rainey and You (2001) studied the effect of boiling surface size and angular orientation for boiling of FC-72 on flat surfaces. They followed the procedure of using plain and microporous painted surfaces similar to that used by Chang and You (1997). The CHF is strongly influenced by the orientation for both plain Cu and microporous surfaces. The CHF decreases with inclination angle and the CHF changes from hydrodynamically controlled to the dryout stage for larger angles. Park and Bergles (1988) also studied size and orientation effects for boiling for R113 on plain Cu surfaces. They observed that heater size does not have any appreciable effect on thermal performance.

Mechanism of boiling on enhanced surfaces has not been fully understood yet. However, some models based on theory do give practically useful knowledge on boiling mechanism. Boiling mechanism can be better understood for structured surfaces than that for the porous surfaces. This is because structured surfaces have well-defined dimensions. Boiling mechanism can be understood following the key concepts like:

- Boiling requires re-entrant or doubly re-entrant tunnel-type cavities.
- Subsurface pores on the enhanced surface are substantially larger than those occurring on natural surfaces.
- This reduces the superheat requirements.
- Liquid is supplied via the surface pores to subsurface capillary passages.
- Because of the surface temperature greater than the saturation temperature, thin film evaporation occurs from liquid–vapour interfaces.
- In the subsurface, capillaries are not subjected to the cold liquid temperature existing above the surface.
- The subsurface capillaries are interconnected.
- Vapour in the subsurface structure at one site activates other sites.

Ayub and Bergles (1987, 1990) boiled R113 and water. They have found optimum gap spacing at the fin tip. Nakayama et al. (1980b) observed the optimum pore diameter that varies with saturation pressure. Sathyabhama (2015) carried out experimental study on the enhancement of pool boiling heat transfer from grooved surface. They observed increased heat transfer rates with increase in groove depth.

Fig. 2.17 Schematic of apparatus used by Nakayama et al. (1980a) to view boiling in tunnels. (**b**) Liquid–vapour interface observed in tunnel by Nakayama et al. (1980a) (from Nakayama et al. 1980a)

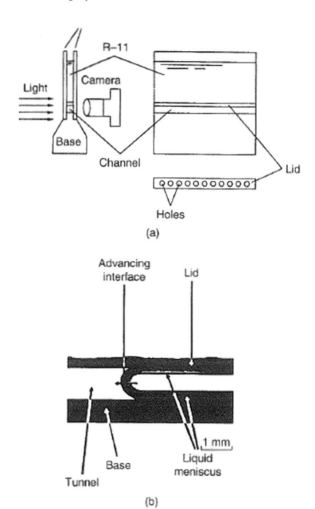

They used Rohsenow correlation to predict their data and observed an error of about 20%. Nakayama et al. (1980a) made a visualization study of boiling in a rectangular tunnel having pores in a very thin copper sheet covering the top surface (Fig. 2.17). Intermittent increase in the wall superheat increased the bubble departure frequency and the number of active pores. Arshad and Thome (1983) visualized boiling using water. However, their experiment was designed to allow observation along the axis of the tunnel, whereas Nakayama et al. (1980a) made viewing from side. Arshad and Thome (1983) concluded that thin film evaporation is the principle of heat transport mechanism inside the tunnel. The geometrical shape of the subsurface groove influences the shape and formation of the evaporating thin liquid film. However, they could not detect a meniscus moving down the tunnel.

Chien and Webb (1998c) visualized boiling on finned tube geometry and closely simulated the actual tube geometry. The vapour-filled tunnel makes the surface force

Fig. 2.18 Boiling
mechanism for a horizontal
tube at low heat flux (from
Chien and Webb 1998c)

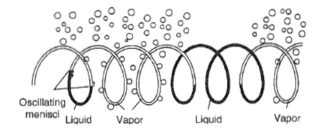

Fig. 2.19 Heat transfer
modes in the tunnels of the
Thermoexcel-E surface
proposed by Nakayama
et al. (1982) (from
Nakayama et al. 1982)

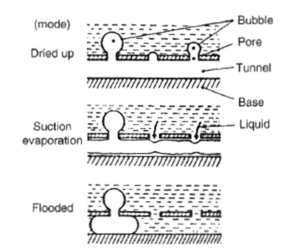

appear as shinning white spots. Liquid filled tunnels are, however, much darker than those on a vapour-filled tunnel, and it is difficult to observe the surface pores. Figure 2.18 shows boiling mechanism for a horizontal tube at low heat flux. This visualization provides the evidence that bubbles are formed by evaporation of menisci in vapour-filled tunnel. Hu and Yeh (2010) developed a new model using Colburn analogy to study film boiling over an isothermal walled sphere, over which liquid nitrogen was made to flow upward. Nakayama et al. (1982) observed three evaporation modes: flooded mode, suction evaporation mode and dried-up mode (Fig. 2.19). Nakayama et al. (1980b) proposed dynamic analytical model to predict boiling performance (Fig. 2.20). They have modelled three different phases of boiling: pressure buildup phase, pressure reduction phase and liquid intake phase. Chien and Webb (1998c) concluded that for saturated boiling an active tunnel is vapour-filled with liquid menisci in the corners.

Table 2.4 summarizes Chien and Webb (1998a, b, 2001) parametric boiling studies. Also, the earlier work of Webb and Pais (1992) observed the boiling phenomenon in the similar way. The high performance with a maximum is observed indicating the dryout heat flux (DHF), and the heat transfer coefficients decrease drastically above the DHF. Above DHF, evaporation does not occur in the tunnels, and the vapour in the tunnels causes high thermal resistance. The boiling curves are

Fig. 2.20 Basis of dynamic model for boiling on the Thermoexcel-E surface proposed by Nakayama et al. (1980b) (from Nakayama et al. 1980b)

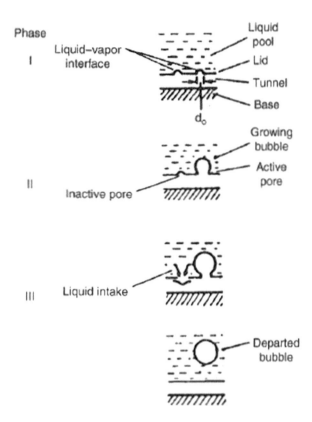

Table 2.4 Tube code and specification tested by Chien and Webb (1998a, b)

Tube code	Fin base shape	Fins/m	P_f (mm)	W_t (mm)	H_f (mm)
1378–0.9	Circular	1378	0.73	0.40	0.90
1378–0.5	Circular	1378	0.73	0.40	0.50
1969–0.9	Circular	1969	0.51	0.25	0.95
1575–0.6	Circular	1575	0.64	0.33	0.60
1575r-0.6	Rectangular	1575	0.64	0.33	0.60

strongly influenced by the pore diameter. Increase in pore diameter causes increase in DHF. Figure 2.21 shows the map showing combined effect of pore diameter and pore pitch. Figures 2.5, 2.6, 2.7 and 2.8 show pool boiling-enhanced surfaces.

Two key geometric characteristics of the pool boiling-enhanced surfaces are (1) substrate tunnels and (2) surface pores or fin gaps. Substrate tunnels have parameters like tunnel pitch, tunnel height, tunnel width, tunnel base radius and tunnel shape. Surface pore geometric parameters are pore pitch and pore diameter. Pore diameter is defined as the diameter of the largest circle that can be drawn within the pore. The models to predict thermal performance of structured surfaces must have the ability to account for the effect of all the geometric dimensions of the boiling surface.

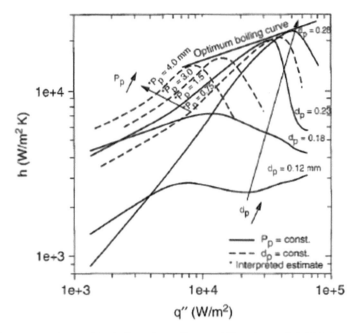

Fig. 2.21 Map showing combined effect of pore diameter (dP) and pore pitch (PP) (from Chien and Webb 1998b)

Empirical correlations do not take into account of the surface and subsurface pore geometries. These correlations consider heat flux and fluid properties for specific enhancement geometry. Zhang and Dong (1992) and Li et al. (1992) have proposed power law correlations for their tube geometry and data. The analytically based model for Nakayama et al. (1980b) considered test surface having subsurface tunnels in a flat plate covered with pores and the vapour bubbles escaped into bulk fluid. The model was based on suction-evaporation mode, and it was assumed that total heat flux was the sum of the latent heat and an external convection single-phase heat flux. Figure 2.22 shows schematic of the surface geometries modelled from (a) Nakayama et al. (1980b) and (b) Xin (1985). Nakayama et al. (1980a) measured the latent heat fraction of the total heat flux for boiling on the enhanced surface using bubble departure diameter and the nucleation site density. The model has several limitations; it requires seven empirical constants; the analysis does not take care of the contact angle directly. Also, the model is limited to suction-evaporation mode.

The Chien and Webb (1998e) model has only two empirical constants, and it accounts for the temporal evaporation rate variation. The model analyses meniscus thickness, bubble departure diameter and bubble growth. It uses bubble frequency and bubble diameter for external heat flux. Chien and Webb (1998a, b, c, d) confirm the Nakayama et al. (1980b) suction and evaporation model for saturated boiling. The Nakayama et al. (1980b) model does not include the bubble frequency and bubble departure diameter in the external convection heat flux. Figure 2.23 shows

Fig. 2.22 Schematic of the surface geometries modelled from (**a**) Nakayama et al. (1980b) and (**b**) Xin (1985)

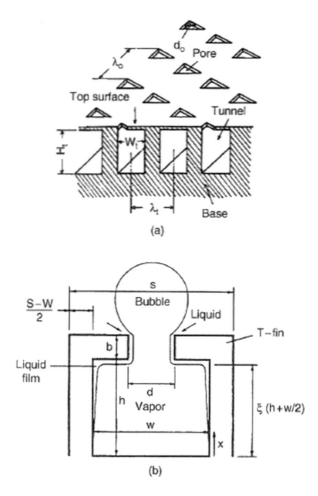

the process of evaporation in the tunnel during a boiling cycle comprising waiting period, bubble growth period and liquid intake period. Waiting period is the period during which liquid is evaporated in the tunnel. But, the vapour is constrained inside the tunnel by the surface tension on the pore, and the pressure in the tunnel increases with time.

In the bubble growth period, vapour passes through surface pores and increases the bubble radius as the bubble grows above the pore. The meniscus radius decreases because the liquid in the tunnel continues to evaporate. Liquid intake period is the period when, after the bubble departure, the pressure in the tunnel is lower than that of the liquid pool and liquid flows into the tunnel and is retained in the corners. The pore diameter and pitch control the amount of liquid that flows into the tunnel. However, in the calculation of bubble frequency, the third and last periods are neglected since the third period is much shorter compared to first and second periods. Figure 2.24 shows a meniscus existing at the top corner of the tunnel. Liquid is

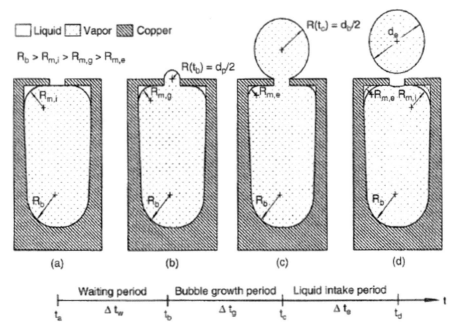

Fig. 2.23 Process of evaporation in the subsurface tunnel for one bubble cycle (from Chien and Webb 1998e)

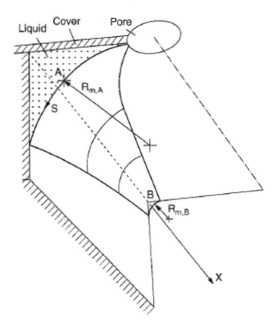

Fig. 2.24 Geometry of the liquid meniscus at the top of the tunnel. Surface tension pulls liquid along the tunnel length (from Chien and Webb 1998e)

drawn into the tunnel by inertia force at the end of a bubble cycle. Surface tension spreads the liquid by pulling it into the corners at the fin base, and evaporation takes place only on the liquid menisci in the corners of the tunnel. One-dimensional model does a fairly good job for the calculation of evaporation rate of the meniscus. Ge et al. (2003) studied the microscale space provided with nano-relief at regular distances for evaporation, thin liquid film and meniscus formation. The impact of nano-roughness on heat transfer augmentation, liquid film formation and the dryout region has been discussed. Xu et al. (2010) studied the sliding of bubble mechanism to enhance boiling heat transfer. They explained that the driving forces in X-direction that lifts the sliding bubble are buoyancy, drag force and mass force, and on the other hand, in the Y-direction the inertial forces at the bubble base was the driving force for the bubble to leave the surface.

The analytically based model of Haider and Webb (1997) as an extension of Mikic and Rohsenow (1969) predicts external heat flux. The model of Mikic and Rohsenow (1969) is based on transient conduction to the superheated liquid layer. The Haider and Webb (1997) model takes the effects of convection to the liquid into consideration. This is due to the convection in the wake of the departing bubbles. Micro-convection, rather than transient conduction, is the principal mode of transport mechanism. Nakayama et al. (1980a) predicted the bubble departure diameter. Chien and Webb (1998d) made a free balance on the bubble using buoyancy and surface tension forces as a function of geometrical parameters, and the prediction does not require any empirical constant. The waiting period and the bubble growth period were modelled separately (Chien and Webb 1998e). The model uses the data of Chien and Webb (1998a, b, 2001). The range of geometric parameters is given in Table 2.5. Ökten and Biyikoglu (2018) reported that due to bubble injection in a thermal storage tank, the heat transfer rates of the storage fluid with heat removal increased by 11% and 14%, respectively.

Ramaswamy et al. (2003) used a polynomial of superheat instead of an empirical constant. Jiang et al. (2001) made some variations in the Khrustalev and Faghri (1994) microheat pipe model, and they assumed that evaporation occurs on liquid meniscus in the tunnel corners. Jiang et al. (2001) obtained the initial meniscus radius by solving the momentum equation for the liquid volume during the intake period. They have accounted for the pressure drop of the liquid flowing through inactive pores. Their model contains an empirical constant which is adjusted to fit the nucleation site data only. Other models due to Xin (1985), Xin and Chao (1987), Wang et al. (1991), Ayub and Bergles (1987) and Webb and Haider (1992) merit good consideration. The models, by and large, are all similar in overall structure, but differ in details only.

Table 2.5 Range of parameters in the database (Webb and Kim 2005)

Fluids	d_p (mm)	P_p (mm)	fins/m	W_t (mm)	H_t (mm)
R11 R123	0.12–0.28	0.75–1.5	1378	0.25–0.4	0.5–1.5
R134a R22	0.18–0.28	0.75–1.5	1578, 1968	0.25–0.33	0.6–1.5

O'Neill et al. (1972) used a thin film concept to understand the boiling mechanism on porous surfaces. They conjectured that vapour bubbles exist within the pores formed by the void space between the stacked particles. There are liquid films on the surface of the particles. Heat conduction across the thin liquid film ensues and evaporation takes place. The interconnected pores in the matrix cause the liquid to be supplied to the pores, and vapour passes through the matrix to the free liquid surface. Pressure builds up in the vapour bubbles with vapour generating within a pore. Vapour finds its way out of the interconnected pores to the liquid surface when the pressure is enough to overcome the surface tension retention force. O'Neill et al. (1972) assumed that each pore contains a vapour bubble, and each pore is an active nucleation site. However, subsequently, Czikk and O'Neill (1979) made a rational assumption that not all pores are active. They observed that the pores can be active, intermittent, liquid-filled and non-functional or closed pores.

However, the modelling attempt, which started with the static model of O'Neill et al. (1972), was further reinforced by more rational dynamic model of Kovalev et al. (1999). Thus, even though initially dynamics of fluid flow within the matrix was not considered and bubble dynamics like bubble departure, bubble frequency and site density remained out of ambit of O'Neill et al. (1972) model; Kovalev et al. model (1999) took into account evaporation from menisci within the matrix and the dynamics of fluid flow within the porous matrix. Figure 2.25 shows the concept and top view of matrix of vapour and liquid-filled channels of Kovalev et al. (1999).

Prediction of porous boiling can be done by using the models of O'Neill et al. (1972) and Kovalev et al. (1999). O'Neill et al. (1972) assumed that thin film evaporation on the particles occurs and Kovalev et al. (1999) assumed that evaporation occurs at menisci within the pore structure. Polezhaev (1990) gives additional information. Nishikawa (1983) developed an empirical correlation of their data on their matric generation (Fig. 2.26). Zhang et al. (1992) has given a power law correlation based on dimensionless parameters assumed to control the boiling process.

CHF on an enhanced surface is higher than that on a plain surface, since the spacing of the vapour jets is different. CHF in nucleate boiling on a plain surface depends on hydrodynamic instability of closely spaced vapour jets which take away latent from the surface (Zuber 1958). Nucleate boiling is not possible in the absence of vapour jets. Lienhard (1987) and Polezhaev (1990) observed that CHF is influenced by the shape of the surface. Cieśliński (2002), Chang and You (1996), Ferjancic and Golobic (2002), Liaw and Dhir (1986), Kandlikar (2001) and Min et al. (2000) give more information on CHF on an enhanced surface.

Faghri (1995), Peterson (1994), Hanlon and Ma (2003) and Ma and Peterson (1997) may be referred for information on thin film evaporation heat transfer mode. Figure 2.27 illustrates the porous sintered wick. A thin porous coating enhances the meniscus evaporation rate (Wang and Catton 2001). Their model has taken care of the fact that thin porous film elongates the width of the evaporating surface area and also improves the capillary force to maintain the liquid supply and increases

Fig. 2.25 (**a**) Concept of
Kovalev et al. (1999) model
for porous boiling surface.
(**b**) Distribution of pressure
in vapour and liquid
channels over matrix depth.
(**c**) Top view of matrix
showing vapour and liquid-
filled channels

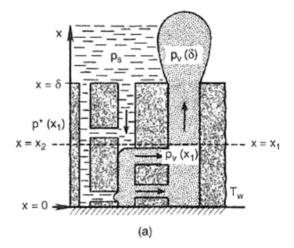

(a)

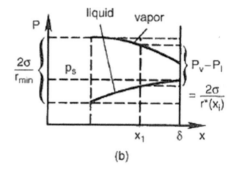

(b)

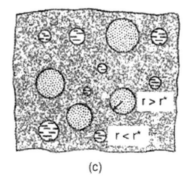

(c)

CHF. Sathyabhama and Pandiyan (2016) have observed the effect of surface
vibration of pool boiling heat transfer enhancement. They observed enhancement
for low frequency and low amplitude of vibration. Also, they used Rohsenow-type
correlation to predict the accuracy of their data and found an average absolute error
of 30%.

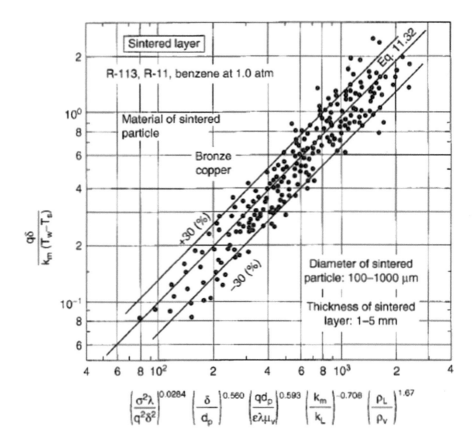

$$\left(\frac{\sigma^2\lambda}{q^2\delta^2}\right)^{0.0284}\left(\frac{\delta}{d_p}\right)^{0.560}\left(\frac{qd_p}{e\lambda\mu_v}\right)^{0.593}\left(\frac{k_m}{k_L}\right)^{-0.708}\left(\frac{\rho_L}{\rho_v}\right)^{1.67}$$

R-113 and R-11 tests conducted using spherical particles

Test	Mat'l.	d_p (μm)	δ (mm)	δ/d_p
1	Cu	250	0.4, 1, 2, 4	1.6, 4, 8, 16
2	Cu	100, 250[a]		
		500	2	4[a], 8[a], 20
3	Bronze	500	1, 2, 3, 4	2, 4, 6, 8
4	Bronze	250[a], 500[a]		
		750, 1000	2	2, 2.7, 4, 8[a]

[a] Also with R-11.

Fig. 2.26 Nishikawa (1983) correlation of their R11 and R113 data for coatings made of uniform spherical particles

Closure The process of nucleate boiling-enhanced surfaces has been described in detail. Porous coatings and re-entrant grooves are exploited in refrigeration industry including process applications. A pore or re-entrant cavity, interconnected cavities and nucleation site having re-entrant shape are responsible for high thermal

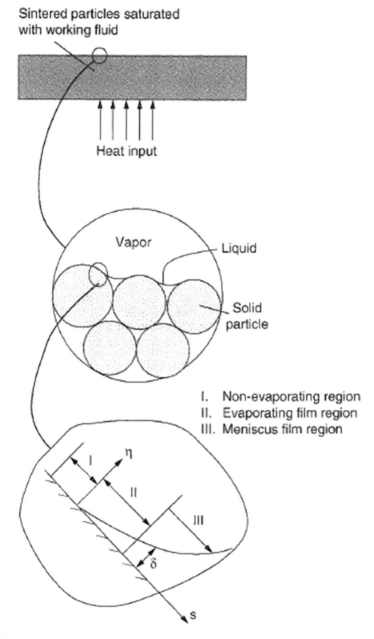

Fig. 2.27 Illustration of porous sintered wick, with liquid menisci existing in the regions between the sintered particles (Hanlon and Ma 2003)

performance. Thin liquid films, active stable vapour trap, low superheat requirement are the advantages. The boiling surface may be less expensive. Some modelling has been done to capture the physics of pool boiling. However, more modelling work is needed. Improved manufacturing method is required.

References

Albertson CE, inventor; Borg-Warner Corp, assignee (1977) Boiling heat transfer surface and method. United States Patent US 4,018,264

Antonelli R, O'Neill PS (1981) Design and application considerations heat exchangers with enhanced boiling surfaces. In: Sourcebook JP (ed) Heat exchanger. Hemisphere, Washington, DC

Arai N (1977) Heat transfer tubes enhancing boiling and condensation in heat exchangers of a refrigerating machine. ASHRAE Trans 83(2):58–69

Arias FJ, Reventos F (2010) Heat transfer enhancement in film boiling due to lift forces on the Taylor-Helmholtz instability in low forced convection from a horizontal surface. J Enhanc Heat Transf 17(2):197

Arshad J, Thome JR (1983) Enhanced boiling surfaces: heat transfer mechanism and mixture boiling. Proc ASME-JSME Therm Eng Joint Conf 1(1):191–197

Asakavicius JP, Zukauskas AA, Gaigolis VA, Eva VK (1979) Heat transfer from freon-113, ethyl alcohol and water with screen wicks. Heat Transf Soviet Res 11(1):92–100

Ayub ZH, Bergles AE (1987) Pool boiling from GEWA surfaces in water and R-113. Wärme-und Stoffübertragung 21(4):209–219

Ayub ZH, Bergles AE (1990) Nucleate pool boiling curve hysteresis for GEWA-T surfaces in saturated R-113. Exp Thermal Fluid Sci 3(2):249–255

Bang IC, Chang SH (2005) Boiling heat transfer performance and phenomena of Al2O3–water nano-fluids from a plain surface in a pool. Int J Heat Mass Transf 48(12):2407–2419

Bar-Cohen A (1992) Hysteresis phenomena at the onset of nucleate boiling. In: Dhir VK, Bergles AE (eds) Pool and external flow boiling. American Society of Mechanical Engineers, New York, pp 1–4

Bell KJ, Mueller AC (1984) Engineering data book II. Wolverine Tube

Bergles AE, Bakhru N, Shires JW (1968) Cooling of high-power-density computer components. MIT Heat Transfer Laboratory, Cambridge, MS

Bergles AE, Chyu MC (1982) Characteristics of nucleate pool boiling from porous metallic coatings. J Heat Transf 104(2):279–285

Bi J et al (2015) Heat transfer characteristics and CHF prediction in nano-fluid boiling. Int J Heat Mass Transf 80:256–265

Bliss FE Jr, Hsu ST, Crawford M (1969) An investigation into the effects of various platings on the film coefficient during nucleate boiling from horizontal tubes. Int J Heat Mass Transf 12 (9):1061–1072

Bolukbasi A, Ciloglu D (2011) Pool boiling heat transfer characteristics of vertical cylinder quenched by SiO2–water nanofluids. Int J Therm Sci 50(6):1013–1021

Bonilla CF, Grady JJ, Avery GW (1965) Pool boiling heat transfer from scored surfaces. Chem Eng Progress Symp Ser 61(57):280–288

Brothers WS, Kallfelz AJ (1979) Heat transfer surface and method of manufacture. US Patent 4,159,739

Carey VP (1992) Liquid-vapor phase-change phenomena. Hemisphere, Washington, DC

Chang JY, You SM (1996) Heater orientation effects on pool boiling of micro-porous-enhanced surfaces in saturated FC-72. J Heat Transf 118(4):937–943

Chang JY, You SM (1997) Boiling heat transfer phenomena from microporous and porous surfaces in saturated FC-72. Int J Heat Mass Transf 40(18):4437–4447

Chaudhri IH, McDougall IR (1969) Ageing studies in nucleate pool boiling of isopropyl acetate and perchloroethylene. Int J Heat Mass Transf 12(6):681–688

Chien L-H, Chang C-C (2003) Enhancement of pool boiling on structured surfaces using HFC-4310 and water. J Enhanc Heat Transf 11:23–44

Chien LH, Chang CC (2004) Enhancement of pool boiling on structured surfaces using HFC-4310 and water. J Enhanc Heat Transf 11(1):23

Chien LH, Chen CL (2000) An experimental study of boiling enhancement in a small boiler. In: 2000 national heat transfer conference, Pittsburgh, PA

Chien LH, Hwang HL (2012) An experimental study of boiling heat transfer enhancement of mesh-on-fin tubes. J Enhanc Heat Transf 19(1):75

Chien LH, Webb RL (1998a) A parametric study of nucleate boiling on structured surfaces, part I: effect of tunnel dimensions. J Heat Transf 120(4):1042–1048

Chien LH, Webb RL (1998b) A parametric study of nucleate boiling on structured surfaces, part II: effect of pore diameter and pore pitch. J Heat Transf 120(4):1049–1054

Chien LH, Webb RL (1998c) Visualization of pool boiling on enhanced surfaces. Exp Thermal Fluid Sci 16(4):332–341

Chien LH, Webb RL (1998d) Measurement of bubble dynamics on an enhanced boiling surface. Exp Thermal Fluid Sci 16(3):177–186

Chien LH, Webb RL (1998e) A nucleate boiling model for structured enhanced surfaces. Int J Heat Mass Transf 41(14):2183–2195

Chien LH, Webb RL (2001) Effect of geometry and fluid property parameters on performance of tunnel and pore enhanced boiling surfaces. J Enhanc Heat Transf 8(5):329

Chu RC, Moran KP, inventors; International Business Machines Corp, assignee (1977) Method for customizing nucleate boiling heat transfer from electronic units immersed in dielectric coolant. United States Patent US 4,050,507

Cieśliński JT (2002) Nucleate pool boiling on porous metallic coatings. Exp Thermal Fluid Sci 25 (7):557–564

Corman JC, McLaughlin MH (1976) Boiling augmentation with structured surfaces. ASHRAE Trans 82(1):906–918

Czikk AM, O'Neill PS (1979) Correlation of nucleate boiling from porous metal films. In: Advances in enhanced heat transfer. ASME, New York, pp 103–113

Dahl MM, Erb LD, Inventors; Gates Rubber Co, assignee (1976) Liquid heat exchanger interface and method. United States Patent US 3,990,862

Danilova GN, Tikhonov AV (1996) R113 boiling heat transfer modeling on porous metallic matrix surfaces. Int J Heat Fluid Flow 17(1):45–51

Das SK, Putra N, Roetzel W (2003) Pool boiling characteristics of nano-fluids. Int J Heat Mass Transf 46:851–862

Dizon MB, Yang J, Cheung FB, Rempe JL, Suh KY, Kim SB (2004) Effects of surface coating on the critical heat flux for pool boiling from a downward facing surface. J Enhanc Heat Transf 11 (2):133

Dundin VA, Danilova GN, Tikhonov AV (1990) Enhanced heat transfer surfaces for shell-andtube evaporators of refrigerating machines. Refrig Mach Ser XM-7:1–46. (in Russian)

Faghri A (1995) Heat pipe science and technology. Global Digital Press

Ferjancic K, Golobic I (2002) Surface effects on pool boiling CHF. Exp Thermal Fluid Sci 25:565–571

Fritz W (1935) Berechnung des maximalvolumens von dampfblasen. Phys Z 36:379–384

Fujie K, Nakayama W, Kuwahara H, Kakizaki K, Inventors; Hitachi Cable Ltd, Hitachi Ltd, assignee (1977) Heat transfer wall for boiling liquids. United States Patent US 4,060,125

Fujii M, Nishiyama E, Yamanaka G (1979) Nucleate pool boiling heat transfer from micro-porous heating surfaces. In: Chenoweth JM, Kaellis J, Michel JW, Shenkman S (eds) Advances in enhanced heat transfer. ASME, New York, pp 45–51

Gaertner RF, inventor; General Electric Co, assignee (1967) Method and means for increasing the heat transfer coefficient between a wall and boiling liquid. United States Patent US 3,301,314

Ge X, Qu W, Zhang L, Ma T (2003) Evaporation heat transfer of thin liquid film and meniscus in micro capillary and on substrate with Nano relief. J Enhanc Heat Transf 10(2)

Gerardi C, Buongiorno J, Hu LW, McKrell T (2011) Infrared thermometry study of nanofluid pool boiling phenomena. Nanoscale Res Lett 6(1):232

Gottzmann CF, ONeill PS, Minton PE (1973) High-efficiency heat-exchangers. Chem Eng Prog 69 (7):69–75

Gottzmann CF, Wulf JB, O'Neill PS (1971) Theory and application of high performance boiling surfaces to components of absorption cycle air conditioners. In: Proceedings conference national gas research technology session V paper, vol 3

Grant AC, inventor; Union Carbide Corp, assignee (1977) Porous metallic layer and formation. United States Patent US 4,064,914

Griffith P, Wallis JD (1960) The role of surface conditions in nucleate boiling. Chem Eng Prog Symp Ser 56(49):49–63

Guglielmini G, Misale M, Schenone C, Pasquali C, Zappaterra M (1988) On performances of nucleate boiling enhanced surfaces for cooling of high-power electronic devices. In: Proceedings of the 22nd international symposium heat transfer in electronic and microelectronic equipment, pp 589–600

Haider SI, Webb RL (1997) A transient micro-convection model of nucleate pool boiling. Int J Heat Mass Transf 40(15):3675–3688

Hanlon MA, Ma HB (2003) Evaporation heat transfer in sintered porous media. J Heat Transf 125 (4):644–652

Hasegawa S, Echigo R, Irie S (1975) Boiling characteristics and burnout phenomena on heating surface covered with woven screens. J Nucl Sci Technol 12(11):722–724

Hausner HH, Mal MK (1982) Handbook of powder metallurgy. Chemical Pub. Co., New York

Hegde RN, Rao SS, Reddy RP (2012) Experimental studies on CHF enhancement in pool boiling with CuO-water nanofluid. Heat Mass Transf 48(6):1031–1041

Hsieh SS, Weng CJ (1997) Nucleate pool boiling from coated surfaces in saturated R-134a and R-407c. Int J Heat Mass Transf 40(3):519–532

Hsieh SS, Yang TY (2001) Nucleate pool boiling from coated and spirally wrapped tubes in saturated R-134a and R-600a at low and moderate heat flux. J Heat Transf 123(2):257–270

Hsu YY (1962) On the size range of active nucleation cavities on a heating surface. J Heat Transf 84 (3):207–213

Hu HP, Yeh RH (2010) Effects of interfacial shear in forced convection turbulent film boiling on a sphere with upward external flowing liquid. J Enhanc Heat Transf 17(2):125

Hübner P, Künstler W (1997) Pool boiling heat transfer at finned tubes: influence of surface roughness and shape of the fins. Int J Refrig 20(8):575–582

Hummel RL, inventor; Dept of Chemical Engineering, assignee (1965) Means for increasing the heat transfer coefficient between a wall and boiling liquid. United States Patent US 3,207,209

Imadojemu HE, Hong KT, Webb RL (1995) Pool boiling of R-11 refrigerant and water on oxidized enhanced tubes. J Enhanc Heat Transf 2(3):189

Jamialahmadi M, Müller-Steinhagen H (1993) Scale formation during nucleate boiling—a review. Corros Rev 11(1–2):25–54

Janowski KR, Shum MS, Bradley SA, inventors; UOP LLC, assignee (1978) Heat transfer surface. United States Patent US 4,129,181

Jiang YY, Wang WC, Wang D, Wang BX (2001) Boiling heat transfer on machined porous surfaces with structural optimization. Int J Heat Mass Transf 44(2):443–456

Jung JY, Kim ES, Kang YT (2012) Stabilizer effect on CHF and boiling heat transfer coefficient of alumina/water nanofluids. Int J Heat Mass Transf 55(7–8):1941–1946

Jung JY, Kim ES, Nam Y, Kang YT (2013) The study on the critical heat flux and pool boiling heat transfer coefficient of binary nanofluids (H2O/LiBr Al2O3). Int J Refrig 36(3):1056–1061

Kajikawa T, Takazawa H, Mizuki M (1983) Heat transfer performance of metal fiber sintered surfaces. Heat Transfer Eng. 4(1):57–66

Kandlikar SG (2001) A theoretical model to predict pool boiling CHF incorporating effects of contact angle and orientation. J Heat Transf 123(6):1071–1079

Kang MG (2000) Effect of surface roughness on pool boiling heat transfer. Int J Heat Mass Transf 43(22):4073–4085

Kartsounes GT (1975) A study of surface treatment on pool boiling heat transfer in refrigerant-12. ASHRAE Trans 81(Pt. 1):320–326

Kedzierski MA (1995) Calorimetric and visual measurements of R123 pool boiling on four enhanced surfaces, Report # NISTIR 5732, US Department of Energy

Khrustalev D, Faghri A (1994) Thermal analysis of a micro heat pipe. J Heat Transf 116 (1):189–198

Kim CJ, Bergles AE (1988) Incipient boiling behaviour of porous boiling surfaces used for cooling of microelectronic chips. Particul Phenom Multiphase Transport 2:3–18

Kim NH (1996) Pool boiling heat transfer enhancement by perforated plates. American Society of Mechanical Engineers, New York, NY

Kim NH, Choi KK (2001) Nucleate pool boiling on structured enhanced tubes having pores with connecting gaps. Int J Heat Mass Transf 44(1):17–28

Ko SY, Liu L, Yao YQ (1992) Boiling hysteresis on porous metallic coatings. In: Chen XJ, Veziroglu TN, Tien CL (eds) Multiphase flow and heat transfer: second international symposium. Hemisphere, New York, pp 259–268

Kole M, Dey TK (2012) Investigations on the pool boiling heat transfer and critical heat flux of ZnO-ethylene glycol nanofluids. Appl Therm Eng 37:112–119

Komendantov AS, Yang Y, Kuang B, Bolshakov RN (2004) Heat transfer enhancement at boiling crisis in straight and spiral tubes. J Enhanc Heat Transf 11(4)

Kovalev SA, Solovyev S, Ovodkov O (1999) Theory of boiling heat transfer on a capillary porous surface. In: Proceeding of the 9th international heat transfer conference, vol. 2, pp 105–110

Krikkis RN, Sotirchos SV, Razelos P (2003) Multiplicity analysis of pin fins under multiboiling conditions. J Enhanc Heat Transf 10(1):95

Kulenovic R, Mertz R, Groll M (2002) High speed flow visualization of pool boiling from structured tubular heat transfer surfaces. Exp Thermal Fluid Sci 25(7):547–555

Kun LC, Czikk AM (1969) Surface for boiling liquids. U.S. Patent 3,454,081 (Reissued August 21, 1979 Re. 30,077)

Kurihara HM, Myers JE (1960) The effects of superheat and surface roughness on boiling coefficients. AICHE J 6(1):83–91

Kwark SM, Kumar R, Moreno G, Yoo J, You SM (2010) Pool boiling characteristics of low concentration nanofluids. Int J Heat Mass Transf 53(5–6):972–981

Li Z, Tan Y, Wang S (1992) Investigation of the heat transfer performance of mechanically made porous surface tubes with ribbed tunnels. In: Chen XJ, Veziroglu TN, Tien CL (eds) Multiphase flow and heat transfer; second international symposium, vol 1. Hemisphere, New York, pp 700–707

Liang HS, Yang WJ (1998) Nucleate pool boiling heat transfer in a highly wetting liquid on micro-graphite-fiber composite surfaces. Int J Heat Mass Transf 41(13):1993–2001

Liaw SP, Dhir VK (1986) Effect of surface wettability on transition boiling heat transfer from a vertical surface. In: Proceedings of the 8th international heat transfer conference, vol 4, pp 2031–2036

Lienhard JH (1987) A heat transfer textbook, 2nd edn. Prentice-Hall, Englewood Cliffs, NJ

Liu JW, Lee DJ, Su A (2001) Boiling of methanol and HFE-7100 on heated surface covered with a layer of mesh. Int J Heat Mass Transf 44(1):241–246

Liu X, Ma T, Wu J (1987) Effects of porous layer thickness of sintered screen surfaces on pool nucleate boiling heat transfer and hysteresis phenomena. In: Wang B-X (ed) Heat transfer science and technology. Hemisphere, New York, pp 577–583

Liu ZH, Xiong JG, Bao R (2007) Boiling heat transfer characteristics of nanofluids in a flat heat pipe evaporator with micro-grooved heating surface. Int J Multiphase Flow 33(12):1284–1295

Lorenz JJ, Mikic BB, Rohsenow WM, (1974) The effect of surface conditions on nucleate boiling characteristics. In: Proc 5th Int Heat Transfer Conf, vol 4, pp 35–49

Luke A (1997) Pool boiling heat transfer from horizontal tubes with different surface roughness. Int J Refrig 20(8):561–574

Ma HB, Peterson GP (1997) Temperature variation and heat transfer in triangular grooves with an evaporating film. J Thermophys Heat Transfer 11(1):90–97

Ma T, Liu X, Wu J, Li H, (1986) Effects of geometrical shapes and parameters of re-entrant grooves on nucleate pool boiling heat transfer from porous surfaces. In: Heat Transfer 1986, Proc 8th Int Heat Transfer Conf, vol 4, pp 2013–2018

Malyshenko SP, Styrikovich MA (1992) Heat transfer at pool boiling on surfaces with porous coating. In: Chen XJ, Veziroglu TN, Tien CL (eds) Multiphase flow and heat transfer: second international symposium, vol 1. Hemisphere, New York, pp 269–284

Marto PJ, Moulson JA, Maynard MD (1968) Nucleate pool boiling of nitrogen with different surface conditions. J Heat Transf 90(4):437–444

Matijević M, Djurić M, Zavargo Z, Novaković M (1992) Improving heat transfer with pool boiling by covering of heating surface with metallic spheres. Heat Transf Eng 13(3):49–57

Mertz R, Kulenovic R, Chen Y, Groll M (2002) Pool boiling of butane from enhanced evaporator tubes. Heat Transf 3:629–634

Mikic BB, Rohsenow WM (1969) A new correlation of pool-boiling data including the effect of heating surface characteristics. J Heat Transf 91(2):245–250

Milton RM, inventor; Union Carbide Corp, assignee (1968) Heat exchange system. United States Patent US 3,384,154

Milton RM, inventor; Union Carbide Corp, assignee (1970) Heat exchange system. United States Patent US 3,523,577

Milton RM, inventor; Union Carbide Corp, assignee (1971) Heat exchange system with porous boiling layer. United States Patent US 3,587,730

Min J, Webb RL, Bemisderfer CH (2000) Long-term hydraulic performance of dehumidifying heat-exchangers with and without hydrophilic coatings. HVAC&R Res 6(3):257–272

Modahl RJ, Luckeroth VC, inventors; Trane Co, assignee (1982) Heat transfer surface for efficient boiling of liquid R-11 and its equivalents. United States Patent US 4,354,550

Nakayama W, Daikoku T, Kuwahara H, Nakajima T (1980a) Dynamic model of enhanced boiling heat transfer on porous surfaces. Prut I: experimental investigation. J Heat Transf 102:445–450

Nakayama W, Daikoku T, Kuwahara H, Nakajima T (1980b) Dynamic model of enhanced boiling heat transfer on porous surfaces Prut II: analytical modelling. Heat Transf 102:451–456

Nakayama W, Daikoku T, Nakajima T (1982) Effects of pore diameters and system pressure on saturated pool nucleate boiling heat transfer from porous surfaces. J Heat Transf 104(2):286–291

Nishikawa K (1983) Augmented heat transfer by nucleate boiling at prepared surfaces. Proc ASME/JSME Thermal Eng Conf (1):387–393

Nishikawa K, Ito T (1980) Augmentation of nucleate boiling heat transfer by prepared surfaces. In: Heat transfer in energy problems, pp 111–118

O'Connor JP, You SM (1995) A painting technique to enhance pool boiling heat transfer in saturated FC-72. J Heat Transf 117(2):387–393

O'Neill PS, Gottzmann CF, Terbot JW (1972) Novel heat exchanger increases cascade cycle efficiency for natural gas liquefaction. In: Advances in cryogenic engineering. Springer, Boston, MA, pp 420–437

Ökten K, Biyikoglu A (2018) Effect of air bubble injection on the overall heat transfer coefficient. J Enhanc Heat Transf 25(3):195

Orman L (2016) Enhancement of pool boiling heat transfer with pin– fin microstructures. J Enhanc Heat Transf 23(2):137

Pais C, Webb RL (1991) Literature survey of pool boiling on enhanced surfaces. ASHRAE Trans 97(1):79–89

Palm B (1992) Heat transfer enhancement in boiling by aid of perforated metal foils. In: Sunden B, Zukauskas A (eds) Recent advances in heat transfer. Elsevier Science, New York

Park KA, Bergles AE (1988) Effects of size of simulated microelectronic chips on boiling and critical heat flux. J Heat Transf 110(3):728–734

Peterson GP (1994) An introduction to heat pipes. Wiley Interscience, New York

Polezhaev YV (1990) Modelling heat transfer with boiling on porous structures. Therm Eng 37 (12):617–620

Ragi EG, inventor; Union Carbide Corp, assignee (1972) Composite structure for boiling liquids and its formation. United States Patent US 3,684,007

Rainey KN, You SM (2001) Effects of heater size and orientation on pool boiling heat transfer from microporous coated surfaces. Int J Heat Mass Transf 44(14):2589–2599

Ramaswamy C, Joshi Y, Nakayama W, Johnson WB (2003) Semi-analytical model for boiling from enhanced structures. Int J Heat Mass Transf 46(22):4257–4269

Rohsenow WM (1985) Boiling, in handbook of heat transfer fundamentals. McGraw Hill, New York, pp 12–15

Sachar KS, Silvestri VJ, inventors; International Business Machines Corp, assignee (1983) Porous film heat transfer. United States Patent US 4,381,818

Saidi MH, Ohadi M, Souhar M (1999) Enhanced pool boiling of R-123 refrigerant on two selected tubes. Appl Therm Eng 19(8):885–895

Saier M, Kastner HW, Klockler R, inventors; Wieland-Werke AG, assignee (1979) Y and T-finned tubes and methods and apparatus for their making. United States Patent US 4,179,911

Sanborn DF, Holman JL, Ware CD, inventors; Trane Co, assignee (1982) Heat exchange surface with porous coating and subsurface cavities. United States Patent US 4,359,086

Sathyabhama A (2015) Nucleate pool boiling heat transfer from a flat-plate grooved surface. J Enhanc Heat Transf 22(3)

Sathyabhama A, Pandiyan PS (2016) Effect of surface vibration on boiling heat transfer from a copper flat circular disc. J Enhanc Heat Transf 23(4)

Shahmoradi Z, Etesami N, Esfahany MN (2013) Pool boiling characteristics of nanofluid on flat plate based on heater surface analysis. Int Commun Heat Mass Transf 47:113–120

Shum MS, inventor; UOP LLC, assignee (1980) Finned heat transfer tube with porous boiling surface and method for producing same. United States Patent US 4,182,412

Sokol P, Blein P, Gorenflo D, Rott W, Schömann H (1990) Pool boiling heat transfer from plain and finned tubes to propane and propylene. Heat Transf:75–80

Sridharan A, Hochreiter LE, Cheung FB, Webb RL (2002) Effect of chemical cleaning on steam generator tube performance. Heat Transf Eng 23(1):38–47

Srinivasan V, Augustyniak JD, Lockett MJ (2001) Pool boiling experiments with liquid nitrogen on enhanced boiling surfaces. In: Compact heat exchangers and enhancement technology for the process industries-2001: Proceedings of the third international conference on compact heat exchangers and enhancement technology for the process industries held at the Davos Congress Centre, Davos, Switzerland. Begell House Publishers Inc., p 409

Suriyawong A, Wongwises S (2010) Nucleate pool boiling heat transfer characteristics of TiO2–water nanofluids at very low concentrations. Exp Thermal Fluid Sci 34(8):992–999

Szumigala ET (1971) Manufacturing method for boiling surfaces. US Patent 3,566,514

Tarrad AH, Burnside BM (1993) Pool boiling tests on plain and enhanced tubes using a wide-boiling-range mixture. Exp Heat Transf Int J 6(1):83–96

Tatara RA, Payvar P (2000) Pool boiling of pure R134a from a single Turbo-BII-HP tube. Int J Heat Mass Transf 43(12):2233–2236

Thors P, Clevinger NR, Campbell BJ, Tyler JT, inventors; Wolverine Tube Inc., assignee (1997) Heat transfer tubes and methods of fabrication thereof. United States Patent US 5,697,430

Torii T, Hirasawa S, Kuwahara H, Yanagida T, Fujie K (1978) The use of heat exchangers with THERMOEXCEL's tubing in ocean thermal energy power plants. ASME

Tsay JY, Yan YY, Lin TF (1996) Enhancement of pool boiling heat transfer in a horizontal water layer through surface roughness and screen coverage [Erhöhung des Wärmeübergangs beim Behältersieden in einer horizontalen Wasserschicht durch Aufrauhen und/oder Abdecken der Heizfläche mittels eines Edelstahlnetzes]. Heat Mass Transf 32(1–2):17–26

Uhle JL (1998) Boiling heat transfer characteristics of steam generator U-tube fouling. Doctoral dissertation, Massachusetts Institute of Technology

Uma BBK, Rao M, Balikrishnan AR (2000) Enhanced pool boililng heat transfer using interference plates. In: Proceedings of the NHTC '00, 34th national heat transfer conference, pp 911–929

Vachon RI, Nix GH, Tanger GE, Cobb RO (1969) Pool boiling heat transfer from Teflon-coated stainless steel. J Heat Transf 91(3):364–369

Vasiliev LL, Zhuravlyov AS, Shapovalov A (2012) Heat transfer enhancement in mini channels with micro/nano particles deposited on a heat-loaded wall. J Enhanc Heat Transf 19(1):13

Vazquez DM, Kumar R (2013) Surface effects of ribbon heaters on critical heat flux in nanofluid pool boiling. Int Commun Heat Mass Transf 41:1–9

Wang CC, Chang YJ, Shieh WY, Yang CY (1998) Nucleate boiling performance of R-22, R-123, R-134A, R-410A, and R-407C on smooth and enhanced tubes. American Society of Heating, Refrigerating and Air-Conditioning Engineers, Inc., Atlanta, GA

Wang DY, Cheng JG, Zhang HJ (1991) Pool boiling heat transfer from T-finned tubes at atmospheric and super-atmospheric pressures. In: ASME HDT, p 159

Wang J, Catton I (2001) Enhanced evaporation heat transfer in triangular grooves covered with a thin fine porous layer. Appl Therm Eng 21(17):1721–1737

Wasekar VM, Manglik RM (2017) Enhanced heat transfer in nucleate pool boiling of aqueous surfactant and polymeric solutions. J Enhanc Heat Transf 24(1–6):47

Webb RL (1981) The evolution of enhanced surface geometries for nucleate boiling. Heat Transf Eng 2(3–4):46–69

Webb RL (1983) Nucleate boiling on porous coated surfaces. Heat Transf Eng 4(3–4):71–82

Webb RL, Donald Q (2004) Kern lecture award paper: odyssey of the enhanced boiling surface. J Heat Transf 126(6):1051–1059

Webb RL, Haider I (1992) An analytical model for nucleate boiling on enhanced surfaces. In: Dhir VK, Bergles AE (eds) Proceedings of the engineering foundation conference on pool and external flow boiling, Santa Barbara, CA, pp 345–360

Webb RL, Kim NY (2005) Principles of enhanced heat transfer. Taylor and Francis, New York

Webb RL, inventor; Trane Co, assignee (1970) Heat transfer surface which promotes nucleate ebullition. United States Patent US 3,521,708

Webb RL, inventor; Trane Co, assignee (1972) Heat transfer surface having a high boiling heat transfer coefficient. United States Patent US 3,696,861

Webb RL, Pais C (1992) Nucleate pool boiling data for five refrigerants on plain, integral-fin and enhanced tube geometries. Int J Heat Mass Transf 35(8):1893–1904

Wei L, Yuan D, Feng Y, Tang D (2014) Experimental study of bubble growth and flow in small-diameter thermosyphon loops with filling ratios of 90% and 95%. J Enhanc Heat Transf 21(1):63

Wen D, Corr M, Hu X, Lin G (2011) Boiling heat transfer of nanofluids: the effect of heating surface modification. Int J Therm Sci 50(4):480–485

Xin MD (1985) Analysis and experiment of boiling heat transfer on T-shaped finned surfaces. In: AICHE paper 23rd national heat transfer conference, Denver, CO

Xin MD, Chao YD (1987) Analysis and experiment of boiling heat transfer on T-shaped finned surfaces. Chem Eng Commun 50(1–6):185–199

Xu J, Chen B, Wang X (2010) Prediction of sliding bubble velocity and mechanism of sliding bubble motion along the surface. J Enhanc Heat Transf 17(2):111

Yang GW, Liang HS, Yang WJ, Vrable DL (1996) Nucleate pool boiling on micro graphite–copper composite surfaces. J Heat Transf 118(3):792–796

Yilmaz S, Westwater JW (1981) Effect of commercial enhanced surfaces on the boiling heat transfer curve. Adv EnhancHeat Transf 18:73–91

You SM, Kim JH, Kim KH (2003) Effect of nanoparticles on critical heat flux of water in pool boiling heat transfer. Appl Phys Lett 83(16):3374–3376

Young RX, Hummel RL (1965) Improved nucleate boiling heat transfer. Chem Eng Prog Symp Ser 61(59):264–470

Zhang H, Dong L (1992) Analysis and experiment of pool boiling heat transfer from Cit-shaped finned tube above atmospheric pressure. In: Chen XJ, Veziroglu TN, Tien CL (eds) Multiphase flow and heat transfer. Second international symposium, vol I. Hemisphere, New York, pp 384–392

Zhang Y, Zhang H, Chen XJ, Verioglu TN, Tien CL (1992) Boiling heat transfer from a thin powder porous layer at low and moderate heat flux. In: II international symposium on multiphase flow heat transfer, New York. Hemisphere, Washington, DC, pp 358–366

Zhou X, Bier K (1997) Pool boiling heat transfer from a horizontal tube coated with oxide ceramics. Int J Refrig 20(8):552–560

Zohler SR, inventor; Carrier Corp, assignee (1990) Porous coating for enhanced tubes. United States Patent US 4,890,669

Zuber N (1958) On the stability of boiling heat transfer. Trans Am Soc Mech Eng 80

Chapter 3
Flow Boiling Enhancement Techniques

3.1 Introduction, Fundamentals, Flow Patterns, Convection, Pressure Drop and Flow Orientation

In the pool boiling discussed earlier, the vapour generated was removed by the departing bubbles. But for flow boiling, the vapour generated is confined to the wall. Thus, the local heat transfer coefficient gets affected depending upon the flow pattern, the mapping of which is obtained from the boiling itself. The flow pattern changes with vapour quality. Flow stratification is important for flow boiling in horizontal tubes. This is not the case for flow boiling in vertical tubes. The tubes may be circular as well as non-circular. There may be annulus as well. Fundamentals of two-phase flow convective heat transfer and convective vaporization need to be studied. The convection dictates the vaporization or condensation, contrary to pool boiling or vapour condensation without vapour velocity. Vapour quality increases continuously during vaporization and the flow pattern changes from one to the other. The heat transfer coefficient is affected by the flow patterns. Kazachkov and Palm (2005) studied the instabilities in two-phase annular flow with a thin film flow on the channel wall. Wen et al. (2015) studied the flow boiling characteristics of R-600 a flow in an annular tube with perforated porous insert.

Figure 3.1 shows flow patterns for complete vaporization for upward flow in a vertical tube. Figures 3.2 and 3.3 show the flow patterns for complete vaporization in a horizontal tube. Stratification effects are more severe at lower vapour qualities, because lower vapour quality makes vapour shear force small. Vapour may be generated by nucleate boiling at tube wall (Fig. 3.2). Flow pattern as a function of mass velocity and vapour quality has been characterized for a particular flow pattern. Hewitt and Roberts (1969) (Fig. 3.4), Baker (1954) (Fig. 3.5), Taitel and Dukler (1976) and Carey (1992) dealt extensively with flow boiling in a tube. Reverse flow patterns are observed by a condensing flow.

S. K. Saha et al., *Two-Phase Heat Transfer Enhancement*, SpringerBriefs in Applied Sciences and Technology, https://doi.org/10.1007/978-3-030-20755-7_3

Fig. 3.1 Flow patterns for
complete vaporization for
upward flow in a vertical
tube (from Carey 1992)

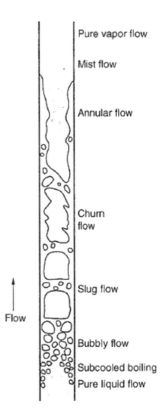

In case of convective vaporization in tubes (Fig. 3.2), nucleate boiling (subcooled, bubbly and plug flow), thin film evaporation (annular flow) and single-phase heat transfer in the gas in the dry wall region downstream of the annular flow region are observed. The dry wall region may or may not exist depending on whether the critical heat flux exceeds in the annular flow region. Liquid velocity tends to suppress nucleate boiling as observed by Chen (1966). In the dry wall region, the heat transfer coefficient will be given by single-phase correlation for heat transfer to a gas. Figure 3.6 shows the convective vaporization curve as a log-log plot of heat flux versus wall superheat. Steiner and Taborek (1992) have observed that an asymptotic model gives a better fit of the data than the superposition model. The total heat transfer coefficient is often quite sensitive to heat flux but relatively insensitive to flow rate. In case of internal fins, the enhancement is more due to convective vaporization than the nucleate boiling, and the total heat transfer coefficient remains to be sensitive to flow rate but insensitive to heat flux. The insert devices mostly enhance the convective term for tube-side enhancements. Capillary forces are used to transport liquid to the upper tube wall when the tube inner surface is flushed with fine grooves at a helix angle. Turbulent mixing of the liquid film is the driving force behind enhancement in tubes with artificial roughness. The roughness provides

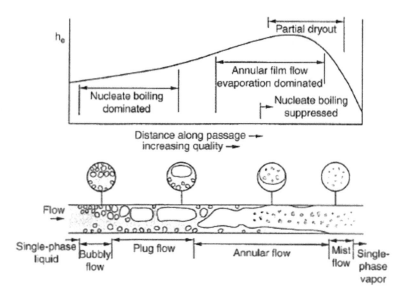

Fig. 3.2 Flow patterns for complete vaporization in a horizontal tube (from Carey 1992)

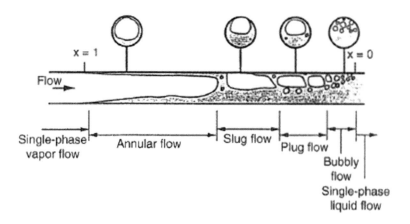

Fig. 3.3 Flow patterns for complete condensation in a horizontal tube (from Carey (1992))

droplet entertainment at moderate vapour qualities. This causes a thinner film on the tube wall. The possibility of low heat transfer coefficient has to be taken into cognition in case of dry wall. Moreover, in case of horizontal tube, gravity forces cause stratification of the flow which in turn causes the upper part of tube dry or intermittently dry. Hwang et al. (2005) studied microfin tubes with single groove with and without tube expansion. The expanded tube showed 16.5% lower heat transfer coefficient and 7.7% higher pressure drop over that of the unexpanded tube.

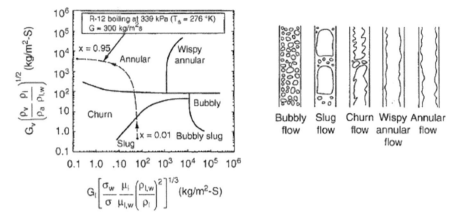

Fig. 3.4 Hewitt and Roberts (1969) flow pattern map for vertical co-current upward flow. The figure is annotated to show the flow patterns encountered evaporating by R-12 (Carey 1992)

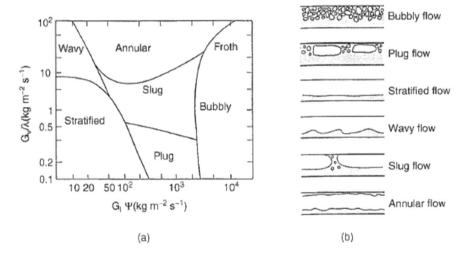

Fig. 3.5 (**a**) Flow pattern map of Baker (1954) for a horizontal tube, (**b**) defined flow patterns

Hu et al. (2011) studied flow boiling heat transfer enhancement in microfin tube for R-410A and oil mixture. They reported 30% enhancement in heat transfer with oil at low vapour qualities (<0.8). At higher vapour qualities, the heat transfer drops with increase in oil condensation.

Fig. 3.6 Superposition
model plotted in the form
q vs. D.Tws (from Webb
and Gupte 1992)

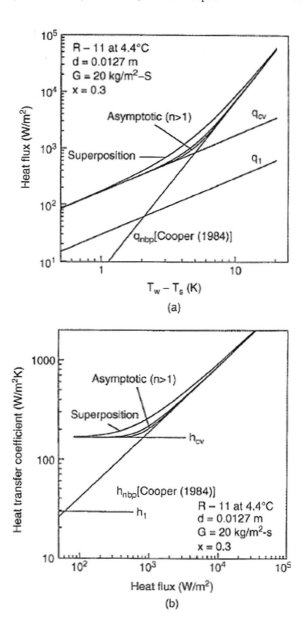

3.2 Two-Phase Pressure Drop, Effect of Flow Orientation, Tube Bundles, Critical Heat Flux

Total pressure drop is the addition of three components: frictional pressure drop, the pressure drop due to gravity and pressure drop due to acceleration. Martinelli parameter takes care of the pressure drop. Chisholm (1967) has given a method in the polynomial form to calculate the Martinelli parameter. Webb et al. (1990) and Ishihara et al. (1980) have given the empirical constants to calculate the Martinelli parameter.

The flow orientation has profound effect on flow pattern. The following situations are important:

1. Vertical tubes with vapour and liquid flowing down
2. Vertical tubes with vapour and liquid flowing up
3. Horizontal tubes with co-current vapour and liquid flow

The detailed discussion may be obtained from Webb and Kim (2005). Pressure drop is an important consideration in evaluating enhancement for vaporization, particularly for high dT and dp. Webb and Gupte (1992) made a thorough survey of models and correlation for flow boiling in tube and tube banks. Bundle orientation and bundle circularity have the influence on flow pattern. Flow pattern for shell-side boiling have been described by Polley et al. (1980).

Depending on the thermal boundary condition, the CHF for flow boiling is also important just in the case of pool boiling. The heat transfer coefficient falls to a lower value and higher temperature, and heat flux may occur. At the time of DNB (Departure from Nucleate Boiling), the heat transfer coefficient drops to a much lower value when nucleate boiling is replaced by film boiling. At moderate or high vapour qualities, the flow pattern is annular with entrained droplets. In the case of vapour dryout, the liquid film on the tube wall dryout and heat transfer coefficient drops. The dryout condition occurs at lower vapour qualities as the mass velocity and local superheat increase. CHF condition may exist in both shell-side and tube-side vaporization.

3.3 Enhancement Techniques

Enhancement geometries must be different in case of flow boiling than those in case of pool boiling because of different flow pattern and convection which alter the vaporization process. Larger internal fins and microfins are the enhancement geometries. Figure 3.7 shows basic types of internally finned tubes. Boling et al. (1953) and Lavin and Young (1965) used integral internal fins (Fig. 3.8). Kubanek and Miletti (1979), Schliinder and Chawla (1967), Pearson and Young (1970), Wen and Hsieh (1995), Panchal et al. (1992), Chen (1966), Ito and Kimura (1979) and Shinohara and Tobe (1985) have worked on internal fins. Figures 3.9, 3.10

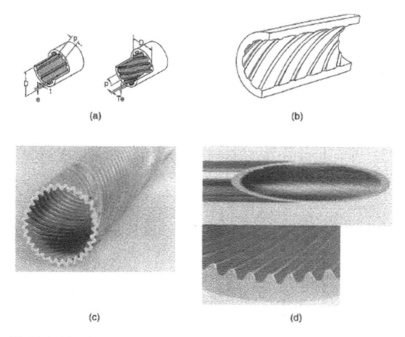

Fig. 3.7 (**a**) Axial and helical internal fins, (**b**) grooved, axial steel tube for power boilers, (**c**) general atomics spirally fluted tube, (**d**) Wieland microfin tube (from Weiland 1991)

and 3.11 give the results and sketches of heat transfer enhancement for flow boiling with internal fins. Parker and Genk (2009) carried out experiment to study the saturation boiling having heat transfer with HFE-7100 on different surfaces having pin surface roughness and inclination.

Swirl flow devices like twisted tape inserts, screw tape inserts, etc. have been used for the enhancement of flow boiling by Bergles et al. (1971), Jensen and Bensler (1986), Agrawal et al. (1982, 1986), Blatt and Adt (1963), Kedzierski and Kedzierski (1997) and Jensen (1985) among others. The twisted tape inserts supress the CHF. Integral roughness such as wire coil insert, corrugated tubes and other techniques was used by Varma et al. (1991), Withers and Habdas (1974), Shinohara and Tobe (1985), Akhanda and James (1991) and Wen and Hsieh (1995). Figures 3.12, 3.13 and 3.14 deal with the integral roughness techniques and their results. Some results and the geometry of the tubes tested are given in Table 3.1 and Table 3.2. Information on heat transfer enhancement for flow boiling in tubes by coatings, perforated foil inserts, porous media and coiled tubes and return bends may be obtained from Czikk et al. (1981), Thome (1990), Chen (1966), Antonelli and O'Neill (1981), Ikeuchi et al. (1984), Wadekar (1998), Palm (1990), Conklin and Vineyard (1992), Topin et al. (1996), Jensen and Bergles (1981), Crain Jr (1973), Campolunghi et al. (1976) and Gu et al. (1989).

It has been observed (Thome 1990) that vapour quality and mass velocity do not have much effect on the high flux tube performance, indicating thereby that nucleate

Fig. 3.8 Tube geometries tested by Lavin and Young (1965) with R-12 and R-22 in vertical and horizontal flow (from Lavin and Young 1965)

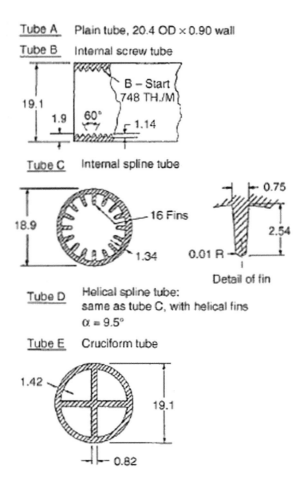

boiling dominates the tube performance. The high flux tube has an effect of mass velocity for horizontal flow (Czikk et al. 1981) because the mass velocity is not high enough to produce annular flow. The upper surface of tube is partially dry, and wetting improves with mass velocity. Even though partial dryout was observed, the high flux performance was still ten times better than that of a plain tube. The high flux tube is superior at all vapour qualities. Wadekar (1998) has observed that heat transfer coefficient is insensitive to the vapour quality both in the high flux tube and in the plain tube, indicating thereby the dominance of nucleate boiling over convective boiling. The best perforated foil insert increases the local heat transfer coefficient by approximately 50%, and the performance is poor at high vapour quality since the space between tube wall and the foil gets dried out. Little information is available for heat transfer enhancement by porous media. Coiled tubes impose centrifugal force on the two-phase flow i.e. a secondary flow is imposed on the two-phase flow regimes, liquid droplets are thrown to the outer wall and the liquid

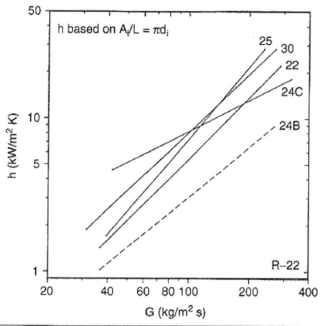

Tube type and number	Number of fins	Tube internal diameter, mm	Tube hydraulic diameter, mm	Fin height, mm	Fin pitch, mm	Wetted area per unit length, mm²/m × 10⁻³	Nominal area per unit length, mm²/m × 10⁻³	Area ratio	Heated length, m
Plain, 24B		14.4	14.4			45.3	45.3	1.00	0.80
Insert, 24C	5	14.4	4.09		610	90.8	45.2	2.00	0.80
Finned, 22	32	14.7	7.57	0.635	305	87.0	46.2	1.88	0.80
Finned, 25	32	14.7	7.57	0.635	152	87.2	46.2	1.89	0.80
Finned, 30	30	11.9	6.30	0.508	102	68.0	37.4	1.82	0.80

Fig. 3.9 R-22 heat transfer vaporization coefficient (h based on A/L = πrd); data of Kubanek and Miletti (1979) at $T = 4.4\,°C$ in 0.8-m-long tube with $\Delta x = 0.7$. The table defines the tube geometries

Fig. 3.10 Tube having five-leg aluminium insert with shrink fit (from Webb and Kim 2005)

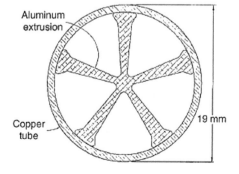

Fig. 3.11 Convective
vaporization coefficient for
R-11 (296–1010 kPa) in
vertical spirally fluted tube
(from Panchal et al. 1992)

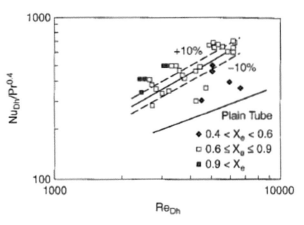

Tube Specifications

Parameter	Value
Tube material	Aluminum, Al 6063
Wall thickness	1.65 mm
Tube flow area	563.2 mm²
Mean inside diameter	26.8 mm
Inside perimeter	137.54 mm
Outside perimeter	136.42 mm
Equivalent diameter	16.38 mm
Flute angle to tube axis	30°
Flute spacing	2.63 mm
Flute depth	1.54 mm
Effective tube length	4.45 m

Fig. 3.12 R-22
vaporization performance
of 19.1-mm-diameter
corrugated tubes of
Withers and Habdas (1974)
plotted vs. severity
factor (Φ)

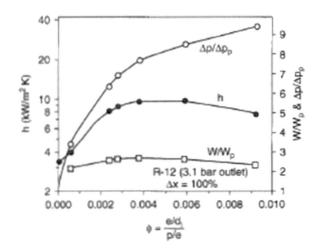

Fig. 3.13 Evaporator tubes
tested by Shinohara and
Tobe (1985): (**a**) star-fin
insert, (**b**) corrugated tube
(CORG), (**c**) microfin
(TFIN), (cl) corrugated
microfin (TFIN-CR) (from
Shinohara and Tobe 1985)

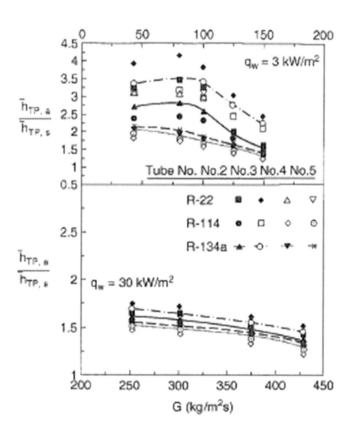

Fig. 3.14 Effect of mass flux (G) on the heat transfer enhancement ratio ($h_{TP,a}/h_{TP,s}$) for the four
tubes tested by Wen and Hsieh (1995) using R-114, R-22 and R-134a

Table 3.1 Tube comparison for R-22 evaporation at $G = 100$ kg/h (Webb and Kim 2005)

Geometry	e_i (mm)	p_i (mm)	h_ik (W/m^2)	Δp (kPa)	h/h_p	$\Delta p/\Delta p_p$
Plain	NA	NA	1.63	20.6	1.00	1.00
Star insert	NA	NA	2.06	35.8	1.26	1.74
CORG	0.7	8.0	2.15	32.4	1.32	1.57
TFIN	NA	NA	2.19	26.0	1.34	1.26
TFIN-CR	0.7	8.0	2.99	32.4	1.83	1.57

NA not applicable

Table 3.2 Tubes tested by Wen and Hsieh (1995) using R-114, R-22 and R-134a

Tube no.	1	2	3	4	5
Tube configuration	Smooth	Star insert (5 legs)	Star insert (10 legs)	Corrugated	Corrugated
Nominal OD (mm)	16.0	16.0	16.0	16.0	16.0
Nominal ID (mm)	14.0	14.0	14.0	14.0	14.0
Nominal heat transfer area (cm^2/m)	440	445	449	471	515
Nominal flow area (cm^2)	6.16	5.81	5.60	7.07	6.17
Helix angle (degree)	–	–	–	30	76
Insert thickness or fin height (mm)	–	1.0	0.8	0.7	0.38
Corrugation pitch (mm)	–	–	–	2.6	1.2

film spirals along the tube wall to the inner surface of the coil due to differential inertia force. This causes an increase of the dryout heat flux. An experimental study on comparison of nucleate boiling heat transfer characteristics of pin fin and straight fin surfaces on an inclined plate using FC-72 working fluid has been carried out by Lee and Chien (2011). As the inclination angle increased, the heat transfer coefficient was found to increase.

3.4 Microfin Tubes

Tatsumi et al. (1982), Shinohara et al. (1987), Fujie et al. (1977), Shinohara and Tobe (1985), Weiland (1991), Torikoshi and Ebisu (1999), Yasuda et al. (1990), Houfuku et al. (2001), Ito and Kimura (1979), Hori and Shinohara (2001) and Tsuchida et al. (1993) were the early researchers who worked on microfin tubes. A plain tube is pulled over a grooved floating plug by a drawing process with rollers applying pressure on the outer tube surface. Figure 3.15 illustrates embossing method in which a flat strip is embossed with the desired pattern, and the tube is rolled to a circular shape and a welded herringbone fin pattern is developed. Single-direction fins or cross-grooved fins may be made by using a second set of rollers that form a secondary embossing pattern. Table 3.3 shows the progress of manufacturing

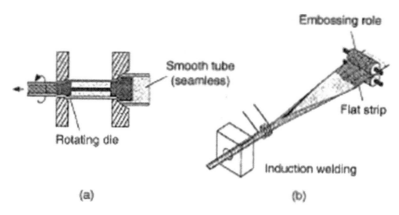

Fig. 3.15 Oxygen boiling performance of high-flux and plain tubes (from Thome 1990)

Table 3.3 Chronological improvement in the Hitachi Thermofin tube (Webb and Kim 2005)

Year	Geometry	d_0	e	p/e	α	β	n	A/A_{ip}	Wt/Wt_p	h/h_p
1977	Orig	9.52	0.15	2.14	25	90	65	1.28	1.22	2.0
1985	EX	9.52	0.20	2.32	18	53	60	1.51	1.19	2.6
1988	HEX	9.52	0.20	2.32	18	40	60	1.60	1.19	3.2
1989	HEX-C	9.52	0.25	2.32	30	40	60	1.73	1.28	3.2
2001	HEX-C	7.0	0.20	2.32	18	40	50			
2001	TFIN-HGL	7.0	0.22	2.03	16	22	54			
2001	HEX-HG	7.0	0.25	2.32	18	15	57			

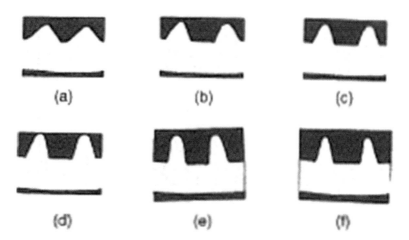

Fig. 3.16 Cross sections of Hitachi Thermofin tubes: (**a**) Thermofin, (**b**) Thermofin EX, (**c**) Thermofin HEX, (**d**) Thermofin HEX-C (from Yasuda et al. 1990). (**e**) TFIN-HG, (**f**) TFIN-HGL (from Houfuku et al. 2001; Hitachi Cable Rev., 20, with permission)

microfins. The fin apex angle, when reduced, reduces the material requirement in the fins. Figure 3.16 shows the cross section of fin geometries. The vaporization coefficient of the thermofin tube is influenced by helix angle. The optimum performance is obtained at a helix angle of approximately 10°. The vaporization coefficient significantly decreases for further increase in helix angle. The heat transfer coefficient increases as the groove pitch increases. Yoshida et al. (1987) have worked on the enhancement mechanism of the microfin tube.

Table 3.4 shows the microfin tube tested by Yoshida et al. (1987). Figure 3.17 shows the circumferentially average R-22 evaporation coefficients in microfin tubes used by Yoshida et al. (1987). Figure 3.18 shows local R-22 evaporation coefficients on top, side and bottom of microfin tubes used by Yoshida et al. (1987). Yoshida et al. (1987) observed that the narrow grooves carry liquid to the sides and top of the tub by capillary wetting, and thin films are provided around the entire tube

Table 3.4 15.8 mm OD microfin tubes tested by Yoshida et al. (1987)

Tube	e	α	N	A/A_p
A	0.24	10	60	1.28
B	0.24	30	60	1.35
C	0.24/0.15	30/15	60	1.35

Fig. 3.17 Circumferentially averaged R-22 evaporation coefficients in 12-mm-ID microfin tubes (from Yoshida et al. 1987)

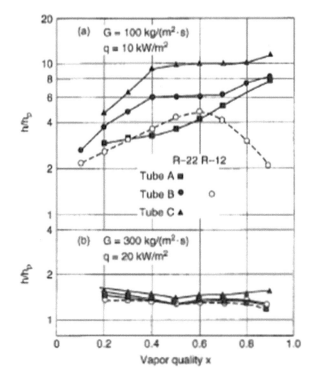

Fig. 3.18 Local R-22
evaporation coefficients on
top, side and bottom of
12-mm-ID microfin tubes
(**a**) $G = 100$ kg/m²-s, (**b**)
$G = 300$ kg/Jn²-s (from
Yoshida et al. 1987)

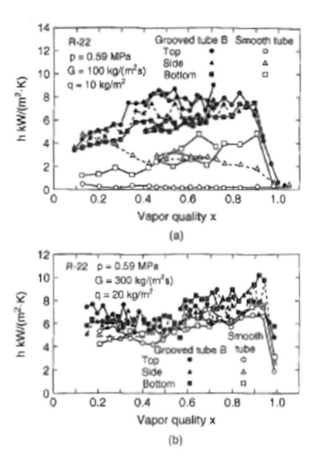

circumference. The dry surface reduces the heat transfer coefficient. There is controversy as to the optimum helix angle is approximately 8°. Schlager et al. (1988a, 1988b) and Eckels and Pate (1991) studied the effect of oil on the thermal performance of several refrigerants, namely R-22, R-134a and R-12. Typical oil concentrations reduce the evaporation coefficient. Ishikawa et al. (2002) investigated the number of fins for best performance. Thome (1996), Newell and Shah (2001), Seo and Kim (2000), Chamra et al. (1996), Muzzio et al. (1998), Ebisu (1999) and Bhatia and Webb (2001) worked on vaporization characteristics in two-dimensional microfin tubes and three-dimensional cross-grooved microfin tube.

Figure 3.19 shows the effect of total inner surface area of the microfin tube on the vaporization coefficient of R-410A. Figures 3.20 and 3.21 shows microfin tubes. The capillary flow in the grooves in two different directions results in a film layer on both the sides of a microfin tube, and mass flux increases which, in turn, enhances the boiling thermal performance. Goto et al. (2001) worked with R-410A vaporization data in herringbone microfin tubes. Kim et al. (2001) used oval microfin tube and compared his results with circular microfin tube. The variation of the installation

Fig. 3.19 Effect of total
inner surface area on the
vaporization coefficient of
R-410A in 7.0-mm-OD
microfin tubes (Hitachi
TFIN-HEX, TFIN-HG,
TFIN-HGL) (from Houfuku
et al. 2001)

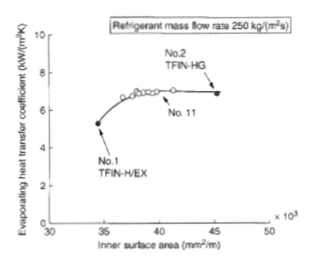

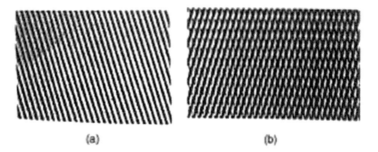

Fig. 3.20 (a) Two-dimensional, (b) three-dimensional cross-grooved 15.88-mm-OD microfin tubes tested by Chamra et al. (1996)

angle of the oval tube made little difference either in heat transfer or in pressure drop. Chamra and Webb (1995) measured local heat transfer coefficient for R-22 for a cross-grooved microfin tube using hot water heating. At higher vapour qualities, when convective effects are prevalent, the heat transfer coefficient for boiling and condensation are nearly equal. So the same mechanism applies for boiling and condensation in the convection dominated regimes. The nucleate boiling compo-nent, condensation and vaporization in a two-dimensional microfin tube have been investigated by Del Col et al. (2002). Vaporization data of Chamra and Webb (1995) and Del Col et al. (2002) are shown in Figs. 3.22 and 3.23, respectively. Zhang et al. (2012) studied flow-jet compound boiling to augment power dissipation in high heat flux electronic devices using microfins on silicon chip. Jones et al. (2009) worked on heat removal from high heat flux chips by using microchannel heat sinks having re-entrant cavities which promote nucleate boiling. They represented their results by using an appropriate homogenous two-phase model. Kim (2015) studied the impact of aspect ratio on the evaporative heat transfer coefficient and friction factor in

Fig. 3.21 Sketches of the groove and wetting pattern of (**a**) two-dimensional, (**b**) herringbone microfin tubes tested by Ebisu (1999)

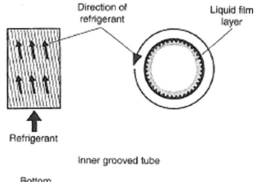

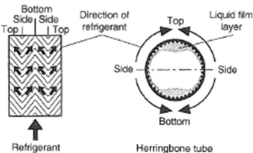

Fig. 3.22 Vaporization heat transfer data compared with the condensation data taken at the same operating condition (R-22 at 24.4 °C) in the 15.88-mm-OD rnicrofin tube (from Chamra and Webb 1995)

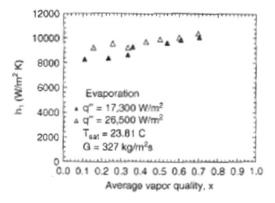

microfin tubes using R-410A as working fluid. He has reported about 1.5–3 enhancement ratio with penalty in pressure drop facing less than 1.

Smaller hydraulic diameter channels are used in electronic cooling. The tubes with inner diameter less than 1 mm (typically few 100 μm) and minichannels (typically greater than 1 mm and less than 3 mm hydraulic diameter) are microchannel and minichannels, respectively, and these are used for manufacturing such channels. These tubes are used for automotive and residential air conditioning application. Kew and Cornwell (1997), Lazarek and Black (1982) and Wambsganss et al. (1993) have worked with minichannels. Their data were for both upflow and

Fig. 3.23 Nucleate pool
data of the two-dimensional
microfin surface (Wolverine
DX-75) taken by Del Col
et al. (2002) using R-22
(from Del Col et al. 2002)

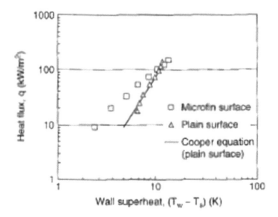

Fig. 3.24 Extruded
aluminium minichannels:
(**a**) 3 × 16 mm,
$D = 1.56$ mm; (**b**)
1.7 × 18 mm, $D = 1.03$ mm;
(**c**) 1.35 × 20 mm,
$D = 0.44$ mm (from Webb
and Kim)

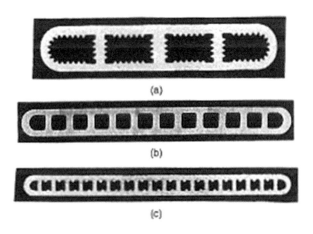

downflow configurations. Relatively high heat flux was applied. Figure 3.24 and
Table 3.5 show the minichannels and studies made there with. Tran et al. (1996) and
Kandlikar (1991) correlated the experimental data and have offered a much useful
correlation used for minichannel. Kasza et al. (1997) made a flow visualization study
in a minichannel and provided insight into the heat transfer coefficient and the
nucleate boiling associated with minichannel. Figure 3.25 shows the generation of
vapour and its progress downstream inside the tube as observed by Kasza et al.
(1997). The flow entered as subcooled liquid and after a threshold value of wall
superheat is attained, the nucleation sites grow bubbles and soon become the size of
the channel cross section, and they get confined into the channel. The bubbles sweep
the channel by the nucleation sites which generate bubbles. The bubbles coalesce
and increase the vaporization rate. The large vapour slug remains in the core of the
channel and a thin liquid film between the wall and slug is formed. The flow
visualization study reveals the increased nucleation frequency and rapid bubble
growth; though the flow dynamics is yet to be fully understood.

Table 3.5 Reported convective vaporization studies in minichannels (Webb and Kim 2005)

References	Tube ID (mm)	Fluid	Operating conditions	Remarks
Lazarek and Black (1982)	3.1	R-113	q: 8.8–90.75 kW/m^2 G: 50–300 kg/m^2 s	Nucleate boiling dominated
Wambsganss et al. (1993)	2.92	R-113	q: 14–380 kW/m^2 G: 125–750 kg/m^2 s	Nucleate boiling dominated
Tran et al. (1996)	2.46 (circular) 2.40 (rectangular)	R-12	q: 3.6–129 kW/m^2 G: 44–832 kg/m^2 s	Nucleate boiling dominated $T > 2.735\,^\circ$C Convection dominated $T < 2.75\,^\circ$C
Yan and Lin (1998)	2.0 (multitube)	R-134a	q: 5–20 kW/m^2 G: 50–200 kg/m^2 s	Both nucleate boiling and convection effect
Zhao et al. (2000)	0.86 (multichannel)	CO_2	q: 3–23 kW/m^2 G: 100–820 kg/m^2 s	Nucleate boiling dominated
Bao et al. (2000)	1.95	R-11 R-123	q: 5–200 kW/m^2 s G: 50–1800 kg/m^2 s	Nucleate boiling dominated
Lin et al. (2001)	1.0	R-141b	q: 10–1150 kW/m^2 G: 300–2000 kg/m^2 s	Both nucleate boiling and convection effect
Yu et al. (2002)	2.98	Water	G: 50–200 kg/m^2 s	Nucleate boiling dominated
Fujita et al. (2002)	1.12	R-123	q: 5–20 kW/m^2 G: 50–400 kg/m^2 s	Nucleate boiling dominated
Pettersen (2003)	0.81 (multichannel)	CO_2	q: 5–20 kW/m^2 G: 190–570 kg/m^2 s	Nucleate boiling dominated

Fig. 3.25 Sketch of the generation of vapour from the wall for water flow at $G = 21$ kg/m^2 s in a rectangular minichannel (2.5 × 6.0 mm) (from Kasza et al. 1997)

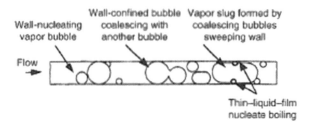

Bao et al. (2000), Yu et al. (2002) and Fujita et al. (2002) may be referred for better understanding of convective vaporization in minichannel. Zhao et al. (2000, 2001) have observed that mass flux does not have appreciable effect on heat transfer coefficient, whereas heat flux makes its influence strongly felt. Zhao et al. (2001) replaced Liu and Winterton (1991) correlation by introducing confinement number to account for the small diameter effect. Pettersen (2003) considered flat tube having rectangular subchannels to investigate convective vaporization of CO_2. Yan and Lin (1998) obtained R-134a convective vaporization data. Webb and Paek (2003) investigated the flow maldistribution in parallel channels. Pettersen (2003) observed significant dryout at high mass flux while using CO_2 in a flat multichannel tube. Mass flux was important for heat transfer at low heat fluxes. Molki et al. (2003) studied the performance of round beads on the walls of minichannel for flow boiling heat transfer enhancement. The results for R-134a have been obtained, and suitable correlations have been proposed. An enhancement of two to eight times that of smooth tube has been observed with their data agreeing with the correlations within 20%. Honda et al. (2003) investigated subcooled nucleate boiling heat transfer from a silicon clip using microfin. Increment in critical heat flux has been observed as 1.9–2.3 times as that of smooth tube. They also commented that the dissolved gas influenced the wall super heat at boiling incipience, while the liquid subcooled heat did not impact much. CHF in convective vaporization is the total heat flux, and it is the sum of the nucleate boiling and connective vaporization contributions. There are two CHF regions. The DNB occurs at subcooled low vapour quality. The high heat flux and dryouts occur at high vapour quality.

Twisted tapes increase DNB and CHF for subcooled boiling (Gambill et al. 1960). Gambill (1963, 1965) has given the reason for increased CHF. Bergles et al. (1971) investigated dryout for nitrogen. They have developed a superposition-based correlation in the evaporation and drywall regions. Cumo et al. (1974) worked with R-12 in vertical tubes. Jensen (1984) developed a correlation to predict the CHF in small diameter tube with twisted tape. Small twist ratio for twisted tape makes significant improvement, whereas for larger twist ratios, the CHF is a weak function of twist ratio. Moreover, thermal insulating effect makes the CHF value even lower than that for plain tubes. Additional data may be obtained from Kabata et al. (1996), Weisman et al. (1994), Weisman and Pei (1983), Weisman and Illeslamlou (1988) and Manglik and Bergles (1993). Grooved tubes have been investigated by Swenson et al. (1962), Weisman et al. (1994), Kitto and Weiner (1982), Chen et al. (1992), Celata et al. (1994) and Lan et al. (1997). CHF for the ribbed tube is three times more than that of a plain tube for the same flow condition and orientation. Lan et al. (1997) provide empirical correlations for CHF enhancement ratio in tube with wire coil inserts. Withers and Habdas (1974) used R-12, and they have observed dryout condition at much higher vapour quality in the corrugated tubes, relative to a plain tube at the same operating conditions. The CHF is increased due to steel mesh and brush type inserts by several times (Mergerlin et al. 1974). However, this was accompanied by much more times of pressure drop increase.

Schlager et al. (1990) reviewed the literature of correlation vaporization inside tube. Enhanced tube correlation contains more geometric variables than those in the case of plain tube correlation. Chen (1966) has developed a model giving an additional convective term and a nucleate boiling term. The model adds a convective term and a nucleate boiling term which depends on heat flux. Pierre (1964) also developed a correlation for convective vaporization. To the contrary of the Chen model (Chen (1966), Pierre (1964) correlation does not contain heat flux. In case of significant nucleate boiling, the Pierre correlation loses the ability to predict data faithfully. In case of absence of nucleate boiling contribution, a correlation in the line of Pierre correlation would be more acceptable. The Azer and Sivakumar (1984) correlation is based on the work of Pierre (1964). This Pierre (1964) equation is often multiplied by the geometry factors to account for the internal fin geometry. Carnavos (1980) correlated single-phase flow data in internally fin tube. Because the correlation was based only on R-113 data, this correlation has limited use. Schliinder and Chawla (1967) developed empirical correlation for aluminium inserts based on their R-11 test data. However, this correlation measurably failed to correlate other data with different fluids.

Recently many predictive methods have been developed for the microfin tube (Cui et al. 1992; Pierre 1964; Koyama et al. 1995; Takamatsu et al. 1993; Thome et al. 1999; Kattan et al. 1998; Cooper 1984; Ravigururajan and Bergles 1985; Brognaux et al. 1997; Cavallini et al. 1998, 2000). All these works have used one or the other refrigerant as working fluid. However, no correlations do a fairly good job for a wide range of data, and the correlations are at variance with others. Nevertheless, the researchers have tried their best so that the correlations do predict at least their own data rather well. Agrawal et al. (1986), Withers and Habdas (1974), Ikeuchi et al. (1984), Chen (1966), Fujita et al. (1986), Muller (1986), Nakajima and Shiozawa (1975), Webb and Gupte (1992), Gupte and Webb (1994, 1995a, b), Webb and Apparao (1990), Jensen and Hsu (1987), Gupte and Webb (1992), Jensen et al. (1992), Yilmaz et al. (1981), Arai et al. (1977), Czikk et al. (1981), Bukin et al. (1982), Memory et al. (1994, 1995), Czikk et al. (1970), Hsieh et al. (2003), Kim et al. (2002), Tatara and Payvar (1999, 2000a, b), Gan et al. (1993) and Marvillet (1989) have dealt with convective vaporization effect in the tube bundles. Figures 3.26, 3.27, 3.28 and 3.29 show the results of works of researchers who dealt with convective vaporization in tube bundles. The effect of spacing between tubes in tube bundles has been studied by Liu and Qiu (2002) and Liu and Tong (2002). Nasr and Behfar (2012) developed an algorithm which considers pressure drop, to study the performance of enhanced tube bundles in evaporative fluid coolers. The effect of

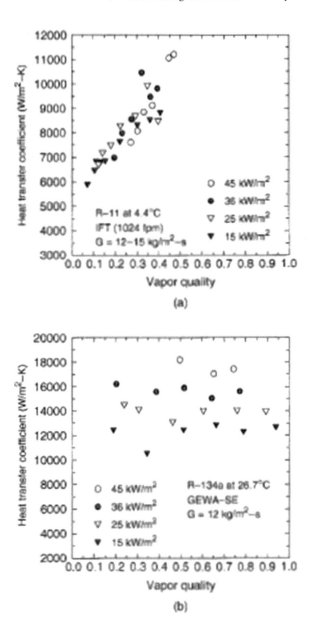

Fig. 3.26 Comparison of 19-mm-0D, 1024 fins/m integral fin and GEWA-SE tubes in a simulated bundle with 23.8-mm equilateral pitch: (**a**) 1024 fins/m integral fin tube with R-11 at 4.4 °C, (**b**) GEWA-SE tube with R-134a at 26.7 °C (from Webb and Kim 2005)

acoustic excitation on boiling heat transfer in tubes immersed in a porous medium has been presented by Zhou et al. (2004). They reported that under acoustic excitation, the results for heat transfer with and without porous medium were found to follow the same trend independent of fluid subcooling. The subcooled flow boiling characteristics of refrigerant R-134a have been experimentally studied in a helically coiled tube. They have presented correlation for heat flux at the onset of

Fig. 3.27 Ratio of pool
boiling-to-convective
vaporization for the
simulated tube bundle data:
(**a**) 1024 fins/m integral fin
with R-11, (**b**) GEWA-SE
tube with R-134a at 26.7 °C
(from Webb and Kim 2005)

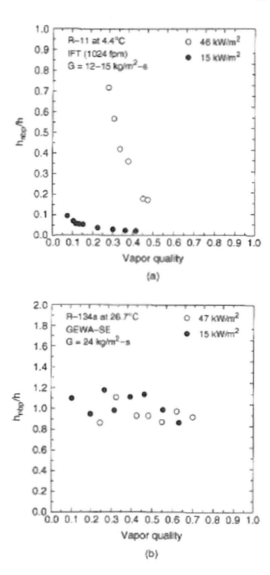

nucleate boiling and also for heat transfer coefficient during nucleate boiling in helically coiled tubes placed in horizontal position.

The models for flow boiling-enhanced surfaces have been developed by Gupte and Webb (1994, 1995a, b), Webb and Chien (1994), Roser et al. (1999) and Kim et al. (2002). The superposition and the asymptotic model have been developed to correlate convective vaporization data on staggered tube banks. Bennett and Chen (1980) modified two-phase multiplier expression for the tube bank configuration. Lewis and Sather (1978) boiled subcooled ammonia feed from a high-flux tube bundle. They have observed that a shutdown causes apparent deactivation of the

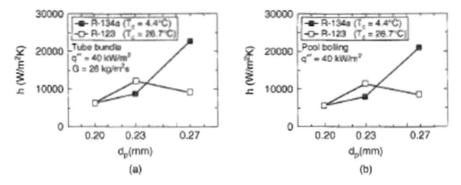

Fig. 3.28 Boiling results on Turbo-B bundle obtained by Kim et al. (2002) for different pore sizes: (**a**) tube bundle, (**b**) single-tube pool boiling

Fig. 3.29 Effect of small gap between tubes for a 10% saltwater solution boiling in a horizontal bundle (7 rows) on 18-mm-diameter enhanced tubes (from Liu and Qiu 2002)

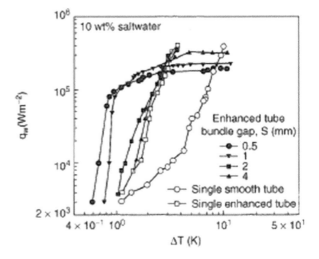

nucleation site resulting from flooding. The starting hysteresis effect was studied by Bergles and Chyu (1982). They have observed that the temperature was not enough to provide sufficient liquid superheating to immediately activate all the nucleation sites, and the activation was a gradual process. O'Neill (1981) has observed that the initial performance can be evidenced in process applications of the high flux boiling surface. Bukin et al. (1982) observed that the sintered porous coatings provide the best performance, which are much better than that of plain or integral fin tubes.

Brazed aluminium plate fin exchanger is used in cryogenic gas processing operations in air separation plants and natural gas liquefaction (Fig. 3.30). Nunez et al. (1999), Robertson (1980), Carey and Shah (1988), Thome (1990), Carey (1992), Mandrusiak and Carey (1989), Kun and Czikk (1969), Thonon et al. (1995), Cornwell and Scoones (1988), Gorenflo (2001), Quazia (2001), Yan and Lin (1999) and Hsieh and Lin (2002) give more valuable practical information on plate fin heat exchangers and the associated convective vaporization on enhanced

Fig. 3.30 Brazed
aluminium plate-and-fin
heat exchanger for
cryogenic service (courtesy
of Altec, Lacrosse, WI)

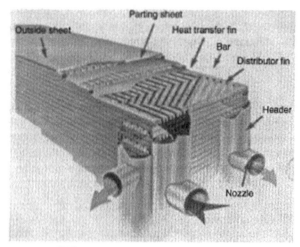

- ASME design pressure—Vacuum to 1,200 psig
- ASME design temperatures—450 to + 150 F
- Cross section—Up to 45" × 72"
- Length—Up to 240"
- Materials—Aluminum alloys 3003, 3004, 5083, 6061

surfaces. Horizontal tubes and vertical tubes behave differently for convective vaporization. Chyu and Bergles (1985), Owens (1978), Conti (1978), Lorenz and Yung (1979), Parken et al. (1990), Rifert et al. (1992), Fujita (1998), Moeykens et al. (1995a, b), Moeykens and Pate (1996), Zeng et al. (1998, 2000), Liu and Yi (2002) and Chyu and Bergles (1989) deal extensively with convective vaporization enhancement on horizontal tubes. It has been observed by the researchers that the fouling film evaporation coefficient of a Turbo-II bundle is 100% larger than the nucleate boiling performance of the Turbo-B bundle. Oil has a pronounced effect on the spray evaporation on enhanced tube bundle. The oil promotes foaming and enhances the heat transfer by returning the formation of dry patches on the bundles. The heat transfer coefficient decreases as the row number in case of tube bundles increases. However, the tube performance is least affected by tube pitch, the drainage pattern on horizontal tubes in a bundle affects the evaporation rate. In this connection, boiling and condensation show the same behaviour as far as the drainage phenomenon is concerned. Yu et al. (2004) studied electrohydrodynamic heat transfer for boiling in horizontal tube bundles and reported that the enhancement in tube bundles was greater as compared to that in a single tube. Also, as the heat flux increased, the EHD enhancement was found to decrease. Huang et al. (2004) also studied the effect of EHD on boiling heat transfer for brass plate in the working fluid.

Chun and Seban (1971), Kosky (1971), Fagerholm et al. (1985), Nakayama et al. (1982), Takahashi et al. (1990), Rifert et al. (1975), Mailen (1980), Chen et al. (2000), Fujita (1999), Megerlin et al. (1974), Ouazia (2001), Thome et al. (1997),

Yasunobu et al. (2002), Zeng & Chyu MC Ayub (2000) and Grimley et al. (1987) have investigated convective vaporization enhancement on vertical tubes. Shatto and Peterson (2017) critically reviewed the performance of twisted tapes to enhance flow boiling heat transfer and developed correlations based on experimental data available in literature. They have also theoretically developed relationship between swirl flow boiling and axial flow boiling. Dong et al. (2019) felt the need to understand the effect of surface modifications of engine cooling passage on subcooled boiling heat transfer enhancement. A 24% increase in heat flux with optimized cavity geometry has been observed. Also, the impact of cavity depth on heat transfer was profound as compared to that of cavity diameter and cavity spacing. Cui et al. (2016) studied the performance of microporous surface to enhance the subcooled flow boiling heat transfer. The microporous surface was made by using a two-step electrodeposition method. They observed that at higher flow rates and higher rates of subcooling, the enhancement rates were low. Mehendale (2016) elaborated the importance of microfin deformation when they are expanded for refrigeration and air conditioning applications. They have concluded that for 10% deformation, there was a fall in flow boiling heat transfer rate by 5%. They have also observed that as the tube cross-sectional area and surface area have been varied due to expansion, the pressure drop decreased by almost 6%. Abdul and Wang (2016) have also worked on the augmentation of flow boiling heat transfer in microfin tubes. They observed that R-1234ze showed higher heat transfer rate and also pressure drop as compared to those of R-134a.

Closure Convective vaporization involves both nucleate boiling and convective evaporation; relative importance between the two modes depends on the surface geometry.

References

Abdul K, Wang CC (2016) Investigation of thermal-hydrodynamic heat transfer performance of R-1234ZE and R-134a refrigerants in a micro-fin and smooth tube. J Enhanc Heat Transf 23 (3):221
Agrawal KN, Varma HK, Lal S (1982) Pressure drop during forced convection boiling of R-12 under swirl flow. J Heat Transf 104(4):758–762
Agrawal KN, Varma HK, Lal S (1986) Heat transfer during forced convection boiling of R-12 under swirl flow. J Heat Transf 108(3):567–573
Akhanda MA, James DD (1991) An experimental study of the relative effects of transverse and longitudinal ribbing of the heat transfer surface in forced convective boiling. J Heat Transf 113 (1):209–215
Antonelli R, O'Neill PS (1981) Design and application considerations for heat exchangers with enhanced boiling surfaces. Paper presented at international conference on advances in heat exchangers, Dubrovnik, Yugoslavia
Arai N, Fukushima T, Arai A, Nakajima T, Fujie K, Nakayama Y (1977) Heat transfer tubes enhancing boiling and condensation in heat exchanger of a refrigerating machine. ASHRAE Trans 83:58–70

Azer NZ, Sivakumar V (1984) Enhancement of saturated boiling heat transfer by internally finned tubes. ASHRAE Trans 90(1A):58–73

Baker O (1954), Simultaneous flow of oil and gas: A report on Magnolia's research on two-phase pipeline design, Oil Gas J 26(5):185–195

Bao ZY, Fletcher DF, Haynes BS (2000) Flow boiling heat transfer of Freon R11 and HCFC123 in narrow passages. Int J Heat Mass Transf 43(18):3347–3358

Bennett DL, Chen JC (1980) Forced convective boiling in vertical tubes for saturated pure components and binary mixtures. AICHE J 26(3):454–461

Bergles AE, Chyu MC (1982) Characteristics of nucleate pool boiling from porous metallic coatings. J Heat Transf 104:279–285

Bergles AE, Fuller WD, Hynek SJ (1971) Dispersed flow film boiling of nitrogen with swirl flow. Int J Heat Mass Transf 14(9):1343–1354

Bhatia RS, Webb RL (2001) Numerical study of turbulent flow and heat transfer in micro-fin tubes—part 2, parametric study. J Enhanc Heat Transf 8(5)

Blatt TA, Adt RR (1963) The effects of twisted tape swirl generators on the heat transfer rate and pressure drop of boiling Freon 11 and water. In: ASME Paper No. ASME-63-WA-42

Boling C, Donovan WJ, Decker AS (1953) Heat transfer of evaporating freon with inner-fin tubing. Refrig Eng 61:1338–1340

Brognaux LJ, Webb RL, Chamra LM, Chung BY (1997) Single-phase heat transfer in micro-fin tubes. Int J Heat Mass Transf 40(18):4345–4357

Bukin VG, Danilova GN, Dyundin VA (1982) Heat transfer from Freons in a film flowing over bundles of horizontal tubes that carry a porous coating. Heat Transf Soviet Res 14(2):98–103

Campolunghi F, Cumo M, Ferrari G, Palazzi G (1976) Full scale tests and thermal design of once-through steam generators. In: AIChE paper presented at 16th national heat transfer conference, St. Louis, MO

Carey VP (1992) Liquid-vapor phase-change phenomena: an introduction to the thermodynamics of vaporization and condensation processes in heat transfer equipment. Hemisphere, Washington, DC

Carey VP, Shah RK (1988) Design of compact and enhanced heat exchangers for liquid-vapor phase-change applications. In: Two-phase flow heat exchangers. Springer, Dordrecht, pp 909–968

Carnavos TC (1980) Heat transfer performance of internally finned tubes in turbulent flow. Heat Transf Eng 1(4):32–37

Cavallini A, Del Col D, Doretti L, Longo GA, Rossetto L (1998) Refrigerant vaporisation inside enhanced tubes: a heat transfer model. In Eurotherm seminar, pp 222–231

Cavallini A, Del Col D, Doretti L, Longo GA, Rossetto L (2000) Heat transfer and pressure drop during condensation of refrigerants inside horizontal enhanced tubes. Int J Refrig 23(1):4–25

Celata GP, Cumo M, Mariani A (1994) Enhancement of CHF water subcooled flow boiling in tubes using helically coiled wires. Int J Heat Mass Transf 37(1):53–67

Chamra LM, Webb RL (1995) Condensation and evaporation in micro-fin tubes at equal saturation temperatures. J Enhanc Heat Transf 2(3):219

Chamra LM, Webb RL, Randlett MR (1996) Advanced micro-fin tubes for evaporation. Int J Heat Mass Transf 39(9):1827–1838

Chen JC (1966) Correlation for boiling heat transfer to saturated fluids in convective flow. Indust Eng Chem Process Des Dev 5(3):322–329

Chen T, Luo YS, Zheng J, Bi Q (2000) Boiling heat transfer and frictional pressure drop in internally rebbed tubes at high pressures. In: Symposium on energy engineering in the 21st century, Hong Kong, pp 393–398

Chen T-K, Chen X-Z, Chen X-J (1992) Boiling heat transfer and frictional pressure drop in internally ribbed tubes. In: Chen X-J, Vezirolu TN, Tien CL (eds) Multiphase flow and heat transfer: second international symposium, vol 1. Hemisphere Pub. Corp., New York, pp 621–629

Chisholm D (1967) A theoretical basis for the Lockhart-Martinelli correlation for two-phase flow. Int J Heat Mass Transf 10(12):1767–1778

Chun KR, Seban RA (1971) Heat transfer to evaporating liquid films. J Heat Transf 93:391–396

Chyu MC, Bergles AE (1985) Enhancement of horizontal tube spray film evaporators by structured surfaces. Adv Enhanced Heat Transf 43:39–47

Chyu MC, Bergles AE (1989) Horizontal-tube falling-film evaporation with structured surfaces. J Heat Transf 111(2):518–524

Conklin JC, Vineyard EA (1992) Flow boiling enhancement of R22 and a nonazeotropic mixture of R143a and R124 using perforated foils. Oak Ridge National Lab, Oak Ridge, TN

Conti RJ (1978) Experimental investigation of horizontal tube ammonia film evaporators with small temperature differentials. In: Proceedings of the 5th Ocean Thermal Energy Conversion (OTEC)

Cooper MG (1984) Saturation nucleate pool boiling—a simple correlation. Chem E Symp Ser 86:786

Cornwell K, Scoones DJ (1988) Analysis of low-quality boiling on plain and low-finned tube bundles. Presented at second UK heat transfer conference, vol 1, pp 21–32

Crain Jr B (1973) Forced convection heat transfer to a two-phase mixture of water and steam in a helical coil. Doctoral dissertation, Oklahoma State University

Cui S, Tan Y, Lu Y (1992) Heat transfer and flow resistance of R-502 flow boiling inside horizontal ISF tubes. In: Multiphase flow and heat transfer: second international symposium, vol 1. Hemisphere, New York, pp 662–670

Cui W, Mungai SK, Wilson C, Ma H, Li B (2016) Subcooled flow boiling on a two-step electrodeposited copper porous surface. J Enhanc Heat Transf 23(2):91

Cumo M, Farello GE, Ferrari G, Palazzi G (1974) The influence of twisted tapes in subcritical, once-through vapor generators in counter flow. J Heat Transf 96(3):365–370

Czikk AM, Gottzmann CF, Ragi EG, Withers JG, Habdas EP (1970) Performance of advanced heat transfer tubes in refrigerant-flooded liquid coolers. ASHRAE Trans 76:96–107

Czikk AM, O'Neill PS, Gottzmann CF (1981) Nucleate pool boiling from porous metal films effect of primary variables. Adv Heat Tran 18:109–122

Del Col D, Webb RL, Narayanamurthy R (2002) Heat transfer mechanisms for condensation and vaporization inside a microfin tube. J Enhanc Heat Transf 9(1)

Dong F, Cao T, Hou L, Ni J (2019) Optimization study of artificial cavities on subcooled flow boiling performance of water in a horizontal simulated engine cooling passage. J Enhanc Heat Transf 26(1):37

Ebisu T (1999) Evaporation and condensation heat transfer enhancement for alternative refrigerants used in air-conditioning machines. In: Heat transfer enhancement heat exchangers. Springer, Dordrecht, pp 579–600

Eckels SJ, Pate MB (1991) In-tube evaporation and condensation of refrigerant-lubricant mixtures of HFC-134a and CFC-12. ASHRAE Trans 97:62–70

Fagerholm NE, Kivioja K, Ghazanfari AR, Jaervinen E (1985) Using structured surfaces to enhance heat transfer in falling film flow. NASA STI/Recon Technical Report No. 87

Fujie K, Itoh N, Innami T, Kimura H, Nakayama N, Yanugidi T (1977) Heat transfer pipe. US Patent 4,044,797, assigned to Hitachi Ltd

Fujita Y (1998) Boiling and evaporation of falling film on horizontal tubes and its enhancement on grooved tubes. In: Kakac S, Bergles AE, Mayinger F, Yuncu H (eds) Heat Transfer Enhancement of Heat Exchangers. Kluwer Academic, Dordrecht, 3259–3346

Fujita Y (1999) Boiling and evaporation of falling film on horizontal tubes and its enhancement on grooved tubes. In: Heat transfer enhanced heat exchangers. Springer, Dordrecht, pp 325–346

Fujita Y, Ohta H, Hidaka S, Nishikawa K (1986) Nucleate boiling heat transfer on horizontal tubes in bundles. In: Proceedings of 8th Int Heat Transfer Conf, pp 2131–2136

Fujita Y, Yang Y, Fujita N (2002) Flow boiling heat transfer and pressure drop in uniformly heated small tubes. In: Heat transfer 2002, 12th international heat transfer conference, vol 3, pp 743–748

Gambill WR (1963) Generalized prediction of burnout heat flux for flowing, subcooled, wetting liquids. Chem Eng Prog Symp Ser 59(41):71–87

Gambill WR (1965) Subcooled swirl-flow boiling and burnout with electrically heated twisted Tapes and Zero Wall Flux. J Heat Transf 87(3):342–348

Gambill WR, Bundy RD, Wansbrough RW (1960) Heat transfer, burnout, and pressure drop for water in swirl flow through tubes with internal twisted tapes. Oak Ridge National Lab, Oak Ridge

Gan YP, Chen Q, Yuan XY, Tian SR (1993) An experimental study of nucleate boiling heat transfer from flame spraying surface of tube bundle in R113/R11-oil mixtures. In: Experimental heat transfer, fluid mechanics, and thermodynamics, p 1226

Gorenflo D (2001) State of the art in pool boiling heat transfer of new refrigerants. Int J Refrig 24 (1):6–14

Goto M, Inoue N, Ishiwatari N (2001) Condensation and evaporation heat transfer of R410A inside internally grooved horizontal tubes. Int J Refrig 24(7):628–638

Grimley TA, Mudawwar IA, Incropera FP (1987) Enhancement of boiling heat transfer in falling films. In: Proc. 1987 ASME-JSME Therm. Eng. Joint Conf., vol 3, pp 411–418

Gu CB, Chow LC, Beam JE (1989) Flow boiling in a curved channel. Heat Transf High Energy High Heat Flux Appl 119:25–32

Gupte NS, Webb RL (1992) Convective vaporization of refrigerants in tube banks. ASHRAE Trans 98(Pt. 2):411–424

Gupte NS, Webb RL (1994) Convective vaporization of pure refrigerants in enhanced and integral-fin tube banks. J Enhanc Heat Transf 1(4):351

Gupte NS, Webb RL (1995a) Convective vaporization data for pure refrigerants in Tube Banks. Part II: enhanced tubes. HVAC&R Res 1(1):48–60

Gupte NS, Webb RL (1995b) Convective vaporization data for pure refrigerants in tube banks, part I: integral-finned tubes. Int J HVAC&R Res 1(1):35–47

Hewitt GF, Roberts DN (1969) Studies of two-phase flow patterns by simultaneous X-ray and flash photography. Atomic Energy Research Establishment Harwell, UK

Honda H, Takamatsu H, Wei JJ (2003) Enhanced boiling heat transfer from silicon chips with micro-pin fins immersed in FC-72. J Enhanc Heat Transf 10(2):211

Hori M, Shinohara Y (2001) Internal heat transfer characteristics of small diameter thermofin tubes. Hitachi Cable Rev 10:85–90

Houfuku M, Suzuki Y, Inui K (2001) High performance, light weight thermos-fin tubes for air-conditioners using alternative refrigerants. Hitachi Cable Rev 20:97–100

Hsieh YY, Lin TF (2002) Saturated flow boiling heat transfer and pressure drop of refrigerant R-410A in a vertical plate heat exchanger. Int J Heat Mass Transf 45(5):1033–1044

Hsieh SS, Huang GZ, Tsai HH (2003) Nucleate pool boiling characteristics from coated tube bundles in saturated R-134a. Int J Heat Mass Transf 46(7):1223–1239

Hu H, Ding GL, Huang XL, Deng B, Gao YF (2011) Experimental investigation and correlation of two-phase heat transfer of R410a/oil mixture flow boiling in a 5-mm microfin tube. J Enhanc Heat Transf 18(3):209–220

Huang X, Li RY, Yu HL (2004) Enhancement of boiling heat transfer for R11 and R123 by applying uniform electric field. J Enhanc Heat Transf 11(4):299

Hwang YW, Kim MS, Kim Y (2005) Evaporation heat transfer and pressure drop in micro-fin tubes before and after tube expansion. J Enhanc Heat Transf 12(1):59

Ikeuchi M, Yumikura T, Fujii M, Yamanaka G (1984) Heat-transfer characteristics of an internal microporous tube with refrigerant 22 under evaporating conditions. ASHRAE Trans 90 (1A):196–211

Ishihara K, Palen JW, Taborek J (1980) Critical review of correlations for predicting two-phase flow pressure drop across tube banks. Heat Transf Eng 1(3):23–32

Ishikawa S, Nagahara K, Sukumoda S (2002) Heat transfer and pressure drop during evaporation and condensation of HCFC22 in horizontal copper tubes with many inner fins. J Enhanc Heat Transf 9(1):17

Ito M, Kimura H (1979) Boiling heat transfer and pressure drop in internal spiral-grooved tubes. Bull JSME 22(171):1251–1257

Jensen MK (1984) A correlation for predicting the critical heat flux condition with twisted-tape swirl generators. Int J Heat Mass Transf 27(11):2171–2173

Jensen MK (1985) An evaluation of the effect of twisted-tape swirl generators in two-phase flow heat exchangers. Heat Transf Eng 6(4):19–30

Jensen MK, Bensler HP (1986) Saturated forced-convective boiling heat transfer with twisted-tape inserts. J Heat Transf 108(1):93–99

Jensen MK, Bergles AE (1981) Critical heat flux in helically coiled tubes. J Heat Transf 103 (4):660–666

Jensen MK, Hsu JT 1987 A parametric study of boiling heat transfer in a tube bundle. In: Proceedings of the 1987 ASME-JSME thermal engineering joint conference, vol 3, pp 133–140

Jensen MK, Trewin RR, Bergles AE (1992) Crossflow boiling in enhanced tube bundles. ASME, New York, NY (USA) 220:11–17

Jones RJ, Pate DT, Thiagarajan N, Bhavnani SH (2009) Heat transfer and pressure drop characteristics in dielectric flow in surface-augmented microchannels. J Enhanc Heat Transf 6(3)

Kabata Y, Nakajima R, Shioda K (1996) Enhancement of critical heat flux for subcooled flow boiling of water in tubes with a twisted tape and with a helically coiled wire. In: ICONE-4: Proceedings. Vol 1—Part B: Basic technological advances

Kandlikar SG (1991) Correlating flow boiling heat transfer data in binary systems. Phase Change Heat Transfer, p 159

Kasza KE, Didascalou T, Wambsganss MW (1997) Microscale flow visualization of nucleate boiling in small channels: mechanisms influencing heat transfer. Argonne National Lab., Lemont, IL

Kattan N, Thome JR, Favrat D (1998) Flow boiling in horizontal tubes: part 3—development of a new heat transfer model based on flow pattern. J Heat Transf 120(1):156–165

Kazachkov IV, Palm B (2005) Analysis of annular two-phase flow dynamics under heat transfer conditions. J Enhanc Heat Transf 12(1):37

Kedzierski MA, Kedzierski MA (1997) Convective Boiling and Condensation Heat Transfer with a Twisted-tape Insert for R12, R22, R152a, R134a, R290, R32/R134a, R32/R152a, R290/R134a, R134a/R600a. US Department of Commerce, National Institute of Standards and Technology

Kew PA, Cornwell K (1997) Correlations for the prediction of boiling heat transfer in small-diameter channels. Appl Therm Eng 17(8–10):705–715

Kim NH (2015) Effect of aspect ratio on evaporation heat transfer and pressure drop of R-410A in flattened microfin tubes. J Enhanc Heat Transf 22(3):177

Kim MH, Shin JS, Bullard CW (2001) Heat transfer and pressure drop characteristics during R22 evaporation in an oval microfin tube. J Heat Transf 123(2):301–308

Kim NH, Cho JP, Youn B (2002) Forced convective boiling of pure refrigerants in a bundle of enhanced tubes having pores and connecting gaps. Int J Heat Mass Transf 45(12):2449–2463

Kitto JB, Weiner M (1982) Effects of nonuniform circumferential heating and inclination on critical heat flux in smooth and ribbed bore tubes. In: 7th international heat transfer conference, Munich

Kosky PG (1971) Thin liquid films under simultaneous shear and gravity flows. Int J Heat Mass Transf 14:1220–1223

Koyama S, Yu J, Momoki S, Fujii T, Honda H (1995) Forced convective flow boiling heat transfer of pure refrigerants inside a horizontal microfin tube. In: Proceedings of the convective flow boiling, pp 137–42

Kubanek GR, Miletti DL (1979) Evaporative heat transfer and pressure drop performance of internally-finned tubes with refrigerant 22. J Heat Transf 101(3):447–452

Kun LC, Czikk AM (1969) Surface for boiling liquids. US Patent 3,454,081 (Reissued 1979, Ref. 30,077), assigned to Union Carbide Corp

Lan J, Disimile PJ, Weisman J (1997) Two phase flow patterns and boiling heat transfer in tubes containing helical wire inserts—part II—critical heat flux studies. 4(4)

Lavin JG, Young EH (1965) Heat transfer to evaporating refrigerants in two-phase flow. AICHE J 11(6):1124–1132

Lazarek GM, Black SH (1982) Evaporative heat transfer, pressure drop and critical heat flux in a small vertical tube with R-113. Int J Heat Mass Transf 25(7):945–960

Lee SC, Chien LH (2011) Experimental study of pool boiling on pin-finned and straight-finned surfaces on an inclined plate in FC-72. J Enhanc Heat Transf 18(4):311

Lewis LG, Sather NF (1978) OTEC performance tests of the Union Carbide flooded-bundle evaporator. Argonne National Lab., Lemont, IL

Lin S, Kew PA, Cornwell K (2001) Two-phase heat transfer to a refrigerant in a 1 mm diameter tube. Int J Refrig 24:51–56

Liu ZH, Qiu YH (2002) Enhanced boiling heat transfer in restricted spaces of a compact tube bundle with enhanced tubes. Appl Therm Eng 22(17):1931–1941

Liu ZH, Tong TF (2002) Boiling heat transfer of water and R-11 on horizontally smooth and enhanced tubes enclosed by a concentric outer tube with two horizontal slots. Experimental heat transfer 15(3):161–175

Liu Z, Winterton RH (1991) A general correlation for saturated and subcooled flow boiling in tubes and annuli, based on a nucleate pool boiling equation. Int J Heat Mass Transf 34(11):2759–2766

Liu ZH, Yi J (2002) Falling film evaporation heat transfer of water/salt mixtures from roll-worked enhanced tubes and tube bundle. Appl Therm Eng 22(1):83–95

Lorenz JJ, Yung D (1979) A note on combined boiling and evaporation of liquid films on horizontal tubes. J Heat Transf 101(1):178–180

Mailen GS (1980) Experimental studies of OTEC heat transfer evaporation of ammonia on vertical smooth and fluted tubes (No. CONF-800633-2). Oak Ridge National Lab., Oak Ridge, TN

Mandrusiak GD, Carey VP (1989) Convective boiling in vertical channels with different offset strip fin geometries. J Heat Transf 111(1):156–165

Manglik RM, Bergles AE (1993) Heat transfer and pressure drop correlations for twisted-tape inserts in isothermal tubes: part I—laminar flows. J Heat Transf 115(4):881–889

Marvillet C (1989) Influence of oil on nucleate pool boiling of refrigerants R-12 and R-22 from Porous Layer Tube. In: Proceedings of eighth eurotherm conference, advances in pool boiling heat transfer, Paderborn, Germany, pp 164–168

Megerlin FE, Murphy RW, Bergles AE (1974) Augmentation of heat transfer in tubes by use of mesh and brush inserts. J Heat Transf 96(2):145–151

Mehendale S (2016) The impact of fin deformation on flow boiling heat transfer and pressure drop inmicrofin tubes. J Enhanc Heat Transf 23(3):197

Memory SB, Chilman SV, Marto PJ (1994) Nucleate pool boiling of a TURBO-B bundle in R-113. J Heat Transf 116(3):670–678

Memory SB, Akcasayar N, Eraydin H, Marto PJ (1995) Nucleate pool boiling of R-114 and R-114-oil mixtures from smooth and enhanced surfaces—II. Tube bundles. Int J Heat Mass Transf 38(8):1363–1376

Moeykens, SA, Pate MB (1996) Effect of lubricant on spray evaporation heat transfer performance of R-134a and R-22 in tube bundles (No. CONF-960254-). American Society of Heating, Refrigerating and Air-Conditioning Engineers, Inc., Atlanta, GA

Moeykens SA, Huebsch WW, Pate MB (1995a) Heat transfer of R-134a in single-tube spray evaporation including lubricant effects and enhanced surface results. American Society of Heating, Refrigerating and Air-Conditioning Engineers, Inc., Atlanta, GA

Moeykens SA, Newton BJ, Pate MB (1995b) Effects of surface enhancement, film-feed supply rate, and bundle geometry on spray evaporation heat transfer performance. American Society of Heating, Refrigerating and Air-Conditioning Engineers, Inc., Atlanta, GA

Molki M, Ohadi MM, Rupani AP, Franca FH (2003) Enhanced flow boiling of R-134a in a minichannel plate evaporator. J Enhanc Heat Transf 10(1):1

Muller J (1986) Boiling heat transfer on finned tube bundles—the effect of tube position and intertube spacing. In: Proceedings of the 8th international heat transfer conference, vol 4, pp 2111–2116

Muzzio A, Niro A, Arosio S (1998) Heat transfer and pressure drop during evaporation and condensation of R22 inside 9.52-mm OD microfin tubes of different geometries. J Enhanc Heat Transf 5(1):39

Nakajima K, Shiozawa A (1975) An experimental study on the performance of a flooded type evaporator. Heat Transf Jpn Res 4(3):49–66

Nakayama W, Daikou T, Nakajima T (1982) Enhancement of Boiling and Evaporation on Structured Surfaces with Gravity-Driven Flow of R-11. In: Proceedings of the 7th international heat transfer conference, heat transfer 1982, Munich, Germany, vol 4, pp 409–414

Nasr MR, Behfar R (2012) Enhanced evaporative fluid coolers. J Enhanc Heat Transf 19(2)

Newell TA, Shah RK (2001) An assessment of refrigerant heat transfer, pressure drop, and void fraction effects in microfin tubes. HVAC&R Res 7(2):125–153

Nunez MP, Polley GT, Sunden B, Heggs PJ (1999) Methodology for the design of multi-stream plate-fin heat exchangers. In: Sunden B, Heggs PJ (eds) Recent advances in analysis of heat transfer for fin type surfaces. Computational Mechanics, Billerica, MA, pp 277–293

O'Neill PS (1981) Private communication, February 16. Linde Div., Union Carbide Corp., Tonawanda, NY

Ouazia B (2001) Evaporation heat transfer and pressure drop of HFC-134a inside a plate heat exchanger. In: Proceedings of American Society of Mechanical Engineers

Owens WL (1978) Correlation of thin film evaporation heat transfer coefficient for horizontal tubes. In: Proceedings of the 5th OTEC conference, vol 3, pp 71–89

Palm B (1990) Heat Transfer Augmentation in Flow Boiling by Aid of Perforated Metal Foils. In: ASME paper 90-WNHT-10, ASME, New York

Panchal CB, France DM, Bell KJ (1992) Experimental investigation of single-phase, condensation, and flow boiling heat transfer for a spirally fluted tube. Heat Transf Eng 13(1):42–52

Parken WH, Fletcher LS, Sernas V, Han JC (1990) Heat transfer through falling film evaporation and boiling on horizontal tubes. J Heat Transf 112(3):744–750

Parker JL, El-Genk MS (2009) Saturation boiling of HFE-7100 dielectric liquid on copper surfaces with corner pins at different inclinations. J Enhanc Heat Transf 16(2)

Pearson JF, Young EH (1970) Simulated performance of refrigerant-22 boiling inside of tubes in a four pass shell and tube heat exchanger. AIChE Symp Ser 66(102):164–173

Pettersen J (2003) Two-phase flow pattern, heat transfer, and pressure drop in microchannel vaporization of CO2/discussion. ASHRAE Trans 109:523

Pierre B (1964) Flow resistance with boiling refrigerants: part 1 & part 2. ASHRAE J 6:58–77

Polley GT, Ralston T, Grant ID (1980) Forced crossflow boiling in an ideal in-line tube bundle. In: ASME paper (80-HT), p 46

Quazia B (2001) Evaporation heat transfer and pressure drop of HFC-134A inside a plate heat exchanger. In: Proceedings of the 2001 ASME international mechanical engineering congress and exposition, PID, vol 6, pp 115–123

Ravigururajan TS, Bergles AE (1985) General correlations for pressure drop and heat transfer for single-phase turbulent flow in internally ribbed tubes. In: Augmentation of heat transfer in energy systems, ASME HTD-52, pp 9–20

Rifert VG, Butuzov AI, Belik DN (1975) Heat transfer in vapor generation in a falling film inside a vertical tube with a finely-finned surface. Heat Trans Soviet Res 7(2):22–25

Rifert VG, Putilin J, Podberezny VL (1992) Evaporation heat transfer in liquid films flowing down the horizontal smooth and longitudinally-profiled tubes. In: Institution of Chemical Engineers, Davis Building, Rugby (ENGL)(3-1289)

Robertson JM (1980) Review of boiling, condensing and other aspects of two-phase flow in plate fin heat exchangers. In: Shah RK, McDonald CF, Howard CP (eds) Compact heat exchangers-history, technological advances and mechanical design problems, HTD, vol 10. ASME, New York, pp 17–27

Roser R, Thonon B, Mercier P (1999) Experimental investigations on boiling of n-pentane across a horizontal tube bundle: two-phase flow and heat transfer characteristics. Int J Refrig 22 (7):536–547

Schlager LM, Pate MB, Bergles AE (1988a) Performance of microfin tubes with refrigerant-22 and oil mixtures. ASHRAE J:17–28

Schlager LM, Pate MB, Bergles AE (1988b) Evaporation and condensation of refrigerant-oil mixtures in a smooth tube and a micro-fin tube. ASHRAE Trans 94(3112):149–166

Schlager LM, Pate MB, Bergles AE (1990) Evaporation and condensation heat transfer and pressure drop in horizontal, 12.7-mm microfin tubes with refrigerant 22. J Heat Transf 112 (4):1041–1047

Schliinder EU, Chawla J (1967) Local heat transfer and pressure drop for refrigerants evaporating in horizontal, internally finned tubes. In: Proc Int Cong Refrig paper 2.47

Seo K, Kim Y (2000) Evaporation heat transfer and pressure drop of R-22 in 7 and 9.52 mm smooth/micro-fin tubes. Int J Heat Mass Transf 43(16):2869–2882

Shatto DP, Peterson GP (2017) Flow boiling heat transfer with twisted tape inserts. J Enhanc Heat Transf 24(1–6):21

Shinohara Y, Tobe M (1985) Development of an improved thermofin tube. Hitachi Cable Rev 4:47–50

Shinohara Y, Oizumi K, Itoh Y, Hori M (1987) Inventors Hitachi Cable Ltd, assignee. Heat-transfer tubes with grooved inner surface. United States Patent US 4,658,892

Steiner D, Taborek J (1992) Flow boiling heat transfer of single components in vertical tubes. Heat Transf Eng 13(2):43–69

Swenson HS, Carver JR, Szoeke G (1962) The effects of nucleate boiling versus film boiling on heat transfer in power boiler tubes. J Eng Power 84(4):365–371

Taitel Y, Dukler AE (1976) A model for predicting flow regime transitions in horizontal and near horizontal gas-liquid flow. AICHE J 22(1):47–55

Takahashi K, Daikoku T, Yasuda H, Yamashita T, Zushi S (1990) The evaluation of a falling film evaporator in an R22 chiller unit. ASHRAE Trans 96(2):158–163

Takamatsu H, Momoki S, Fujii T (1993) A correlation for forced convective boiling heat transfer of pure refrigerants in a horizontal smooth tube. Int J Heat Mass Transf 36(13):3351–3360

Tatara RA, Payvar P (1999) Effects of oil on boiling R-123 and R-134a flowing normal to an integral-finned tube bundle. ASHRAE Trans 105:478

Tatara RA, Payvar P (2000a) Effects of oil on boiling of replacement refrigerants flowing normal to a tube bundle-part I. In: ASHRAE Winter Meeting 2000 ASHRAE

Tatara RA, Payvar P (2000b) Effects of oil on boiling of replacement refrigerants flowing normal to a tube bundle—Part II. In: ASHRAE Winter Meeting 2000 ASHRAE

Tatsumi A, Oizumi K, Hayashi M, Ito M (1982) Application of inner groove tubes to air conditioners. Hitachi Rev 32(1):55–60

Thome JR (1990) Enhanced boiling heat transfer. Hemisphere Pub. Corp (Taylor & Francis)

Thome JR (1996) Boiling of new refrigerants: a state-of-the-art review. Int J Refrig 19(7):435–457

Thome JR, Favrat D, Kattan N (1997) Evaporation in microfin tubes: a generalized prediction model

Thome JR, Kattan N, Favrat D (1999) Evaporation in microfin tubes: a generalized prediction model. In: Convective flow and pool boiling. Taylor and Francis, pp 239–244

Thonon B, Vidil R, Marvillet C (1995) Recent research and developments in plate heat exchangers. J Enhanc Heat Transf 2(1–2):149

Topin F, Rahli O, Tadrist L, Pantaloni J (1996) Experimental study of convective boiling in a porous medium: temperature field analysis. J Heat Transf 118(1):230

Torikoshi K, Ebisu T (1999) Japanese advanced technologies of heat exchanger in air-conditioning and refrigeration applications. In: Shah RK, Bell KJ, Honda H, Thonon B (eds) Compact heat exchangers and enhancement technology for the process industries. Begell Honse, New York, pp 17–24

Tran TN, Wambsganss MW, France DM (1996) Small circular-and rectangular-channel boiling with two refrigerants. Int J Multiphase Flow 22(3):485–498

Tsuchida T, Yasuda K, Hori M, Otani T (1993) Internal heat transfer characteristics and workability of narrow thermofin tubes. Hitachi Cable Rev 12:97–100

Varma HK, Agrawal KN, Bansal ML (1991) Heat transfer augmentation by coiled wire turbulence promoters in a horizontal refrigerant-22 evaporator. Presented at winter meeting, American

Society of Heating, Refrigerating, and Air-Conditioning Engineers, New York, NY. ASHRAE Trans 97:359–364

Wadekar VV (1998) A comparative study of in-tube boiling on plain and high flux coated surfaces. J Enhanc Heat Transf 5(4):257

Wambsganss MW, France DM, Jendrzejczyk JA, Tran TN (1993) Boiling heat transfer in a horizontal small-diameter tube. J Heat Transf 115(4):963–972

Webb RL, Apparao TV (1990) Performance of flooded refrigerant evaporators with enhanced tubes. Heat Transf Eng 11(2):29–44

Webb RL, Chien LH (1994) Correlation of convective vaporization on banks of plain tubes using refrigerants. Heat Transf Eng 15(3):57–69

Webb RL, Choi K-D, Apparao T (1990) A theoretical model to predict the heat duty and pressure drop in flooded refrigerant evaporators. ASHRAE Trans 95(Pt. 1):326–338

Webb RL, Gupte NS (1992) A critical review of correlations for convective vaporization in tubes and tube banks. Heat Transf Eng 13(3):58–81

Webb RL, Kim NH (2005) Principles of enhanced heat transfer, 2nd edn. Taylor & Francis

Webb RL, Paek JW (2003) Letter to the editors—concerning paper published by Y.-Y. Yan and T.-F. Lin. Int J Heat Mass Transf 46:1111–1112

Weiland 1991 Ripple fin tubes, Wieland-Werke AG brochure TKI-42e(M)-02.91, Ulm, Germany

Weisman J, Illeslamlou S (1988) A phenomenological model for prediction of critical heat flux under highly subcooled conditions. Fusion Technol 13:654–659

Weisman J, Pei BS (1983) Prediction of critical heat flux in flow boiling at low qualities. Int J Heat Mass Transf 26(10):1463–1477

Weisman J, Yang JY, Usman S (1994) A phenomenological model for boiling heat transfer and the critical heat flux in tubes containing twisted tapes. Int J Heat Mass Transf 37(1):69–80

Wen MY, Hsieh SS (1995) Saturated flow boiling heat transfer in internally spirally knurled/integral finned tubes. J Heat Transf 117(1):245–248

Wen MY, Jang KJ, Ho CY (2015) Flow boiling heat transfer in R-600a flows inside an annular tube with metallic porous inserts. J Enhanc Heat Transf 22(1)

Withers JG, Habdas EP (1974) Heat transfer characteristics of helical-corrugated tubes for intube boiling of refrigerant R-12. AIChE Symp Ser 70:98–106

Yan YY, Lin TF (1998) Evaporation heat transfer and pressure drop of refrigerant R-134a in a small pipe. Int J Heat Mass Transf 41(24):4183–4194

Yan YY, Lin TF (1999) Evaporation heat transfer and pressure drop of refrigerant R-134a in a plate heat exchanger. J Heat Transf 121(1):118–127

Yasuda K, Ohizumi K, Hori M, Kawamata O (1990) Development of condensing thermofin-HEX-C tube. Hitachi Cable Rev:27–30

Yasunobu F, Yang Y, Fujita N (2002) Flow boiling heat transfer and pressure drop in uniformly heated small tubes. Heat Transf 3:743–748

Yilmaz S, Palen JW, Taboerk J (1981) Enhanced boiling surfaces as single tubes and tube bundles. In: Webb RL, Carnavos TC, Park EL, Hostetler KM (eds) Advances in enhanced heat transfer, HTD, vol 18, pp 123–130

Yoshida S, Matsunaga T, Hong HP, Nishikawa K (1987) Heat transfer to refrigerants in horizontal evaporator tubes with internal, spiral grooves. In: Proceedings of the 1987 ASME-JSME thermal engineering joint conference, vol 5, pp 165–172

Yu W, France DM, Wambsganss MW, Hull JR (2002) Two-phase pressure drop, boiling heat transfer, and critical heat flux to water in a small-diameter horizontal tube. Int J Multiphase Flow 28(6):927–941

Yu HL, Li RY, Huang X, Chen ZH (2004) EHD boiling heat transfer enhancement outside horizontal tubes. J Enhanc Heat Transf 11(4):291

Zeng X, Chyu M, Ayub ZH (1998) Ammonia spray evaporation heat transfer performance of single low-fin and corrugated tubes. ASHRAE Trans 104(Pt.1):185–196

Zeng X, Chyu MC Ayub ZH (2000) An experimental study of spray evaporation of Ammonia in a square-pitch, low-fin tube bundle. In: Proceedings of the 34th national heat transfer conference, Pittsburgh, PA. NHTC, pp 2000–12215

Zhang Y, Wei J, Guo D (2012) Enhancement of flow-jet combined boiling heat transfer of FC-72 over micro-pin-finned surfaces. J Enhanc Heat Transf 19(6):489

Zhao Y, Molki M, Ohadi MM, Dessiatoun SV (2000) Flow boiling of CO_2 in microchannels. Univ. of Maryland, College Park, MD

Zhao Y, Molki M, Ohadi MM 2001 Predicting flow boiling heat transfer of C02 microchannels. In: Jaluria Y (ed) Proceedings of the ASME, HTD, Vol 369-3. ASME international mechanical engineering congress and exposition, ASME, New York, pp 195–204

Zhou DW, Liu DY, Cheng P (2004) Boiling heat transfer characteristics from a horizontal tube embedded in a porous medium with acoustic excitation. J Enhanc Heat Transf 11(3):231

Chapter 4
Condensation

4.1 Vapour Space Condensation

Plates and tubes (horizontal as well as vertical orientation) are used for enhanced vapour space condensation. Condensation is enhanced by interfacial shear stress for co-current drainage of condensation film and vapour flow. If the vapour has significant velocity, then the condensation is called convective condensation. Mostly passive types of enhancement techniques are used. Special surface geometries are used for the enhancement of film condensation and non-wetting coatings or additives promote dropwise condensation. Electric field also enhances film condensation. Condensation occurs on a surface with temperature lower than vapour saturation temperature. Condensed liquid on the surface is in the form of a wetted film or in droplets in case of non-wetting surface. Although dropwise condensation gives a very high heat transfer coefficient, after a certain service time, wetting of the surface occurs because of oxidation, and the dropwise condensation is reverted to filmwise condensation due to surface wetting. Surface tension is the driving force for the enhancement of film condensation (Gregorig 1954a, b; Karkhu and Borovkov 1971). Surface tension devices, however, do not increase the surface area of the base surface. Nonetheless, wires, etc. with poor thermal contact may be used. Surface tension helps draining the condensate from the fins and retains condensate within interfin region of finned tubes. Thus, the analytical models for predicting the condensate rate on horizontal integral fin tube have been developed. Also, models have been developed to predict condensate retention and to take the effect of drainage strips into account.

Condensation fundamentals are given in Nusselt (1916). Collier and Thome (1994) have dealt with the more complex situations for condensation.

4.2   Dropwise Condensation

Griffith (1985) has reviewed extensively performance and practical aspects of dropwise condensation. Carey (1992) has dealt with analytical treatment of dropwise condensation. Non-wetting surface coating and chemical additive promote dropwise condensation. Refrigerant or many organic chemicals have low surface tension for which they are not suitable for dropwise condensation. Water with high surface tension does not wet the surface, and steam is a good candidate for dropwise condensation. Iltscheff (1971), Hanneman (1977), Kim et al. (2001), Das et al. (2000a, b), Zhao and Burnside (1994), Ma et al. (2002), Lixin and Jiehui (1998) and Abuorabi (1998) have worked with enhanced dropwise condensation surfaces. Vertical plates and single horizontal tubes rather than tube bundles have drawn more interest by the researchers since the tube bundles are baffled with the problem of inundation and vapour shear effects causing reverting to film condensation. Utaka and Nishikawa (2003) used a high-speed camera to observe the Marangoni dropwise condensation phenomenon with water and ethanol vapour mixture as working fluid. They have used laser light extinction method to calculate the thickness of the condensate film. Yan et al. (2011) discussed the Marangoni condensation of water-ethanol vapour mixture on flat plate, taper fin plate and flat fin plate. They observed that heat transfer enhancement was due to temperature gradients present on the condensing surface. They concluded that the effect of vapour pressure and velocities was negligible under large temperature gradients. Lan et al. (2009) explained the importance of surface free energy difference at the interface of the surface and condensate in dropwise condensation phenomenon. They developed a model that takes into account the surface free energy difference and concluded that the increase in surface free energy difference results in enhanced heat transfer coefficient.

4.3   Enhancement of Film Condensation

Film condensation on vertical plates and tubes and on horizontal tubes may be enhanced by the following techniques: surface coatings, roughness, horizontal integral fin tubes, corrugated tubes, surface tension drainage and electric fields.

Cary and Mikic (1973), Brown and Matin (1971), Glicksman et al. (1973), Notaro (1979), Renken and Aboye (1993a, b), Wang et al. (2000) and Renken and Mueller (1993) have dealt extensively with surface coatings. Figure 4.1 shows the thin condensation films on the particles. Medwell and Nicol (1965), Dipprey and Sabersky (1963) and Nicol and Medwell (1966) are the few who have worked for the enhancement due to a closely knurled roughness for a condensate film.

Investigation with integral fin tubes (Fig. 4.2) has been done by Webb et al. (1985), Wen et al. (1994), Das et al. (1995), Briggs and Rose (1994, 1995), Honda and Makishi (1995), Rudy and Webb (1983, 1985), Honda et al. (1983), Honda and

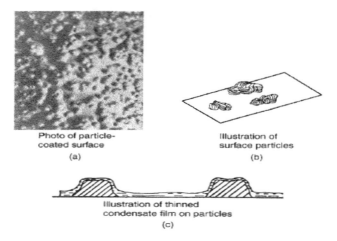

Fig. 4.1 Enhanced condensation surface formed by small diameter metal particles bonded to the base surface. (**a**) Actual surface, (**b**) particles boned to the surface, (**c**) thin condensate on particles and thick condensate film on the base surface (from Webb and Kim 2005)

Fig. 4.2 Horizontal, integral finned tube (from Webb and Kim 2005)

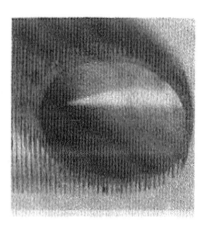

Table 4.1 Enhancement ratio for R-11 condensing on integral-fin tubes ($d_e = 19$ mm, $T_s = 35$ °C, $\Delta T_{vs} = 9.5$ K) (Webb and Kim 2005)

Fins/m (m^{-1})	e_p (mm)	h (mm)	h/h_p (W/m^2-K)
748	1.5	8070	2.64
1024	1.5	11,970	3.91
1378	0.9	16,140	5.28

Nozu (1985, 1987b), Liu et al. (1999), Yau et al. (1986), Sukhatme et al. (1990), Webb and Murawski (1990) and Wang et al. (1990). Tables 4.1 and 4.2 show the investigation on condensation on integral fin tubes.

Table 4.2 Published data for condensation on integral fin tubes (Webb and Kim 2005)

Fins/m	Fluids	Ref.	Comment
608	Methyl chloride, sulphur chloride, R-22 n-pentane, propane	Beatty and Katz (1948)	First model for integral fin tubes
630	R-22, n-butane, acetone, water	Katz and Geist (1948)	Earliest published reference on integral fin tubes
396–770	R-12	Henrici (1961)	
748–1024	R-22	Pearson and Withers (1969)	
748–1024	R-22	Takahashi et al. (1979)	
1060–1610	R-11	Carnavos (1980)	
1020–2000	R-113, methanol	Honda et al. (1983)	
95–1000	Water	Yau et al. (1985)	
100–667	Water	Wanniarachchi et al. (1986)	
400–1000	R-113	Masuda and Rose (1987)	
200–1333	Water	Marto et al. (1988)	
1417	R-11	Sukhatme et al. (1990)	
400–067	R-113	Briggs et al. (1992)	
1000	R-113	Wang et al. (1990)	
1026	R-113	Honda et al. (1994)	Effect of fin tip, mid fin root radius; experimental and prediction
500, 667	R-113, steam, ethylene glycol	Briggs and Rose (1994)	Effect of fin root radius
1333, 1575	R-113	Cavallini et al. (1994)	Visualization of drainage on circular integral fin and spine fin with vapour velocity

(continued)

Table 4.2 (continued)

Fins/m	Fluids	Ref.	Comment
787, 1212, 1490	R-152A	Cheng and Tao (1994)	Circular and spine integral fins
667	R-113	Briggs et al. (1995)	Effect of fin height and cube thermal conductivity
1575, 2000	R-11, R-113	Cavallini et al. (1995)	Refrigeration condensation data
500, 664	Steam	Das et al. (1995)	Effect of fin height and tube thermal conductivity
1333	R-123	Honda and Makishi (1995)	Effect of circumferential ribs on fin side; theoretical prediction
450–1923	R-113	Briggs and Rose (1995)	Data on 12 commercial integral fin tubes; includes T- and Y-shaped fins
200–2000	R-11, R-113, steam	Cavallini et al. (1996)	Model for effect of vapour velocity
300–550	Steam	Jaber and Webb (1996)	Effect of fin geometry, fin pitch and tube thermal conductivity
1024–1640	R-22	Cheng et al. (1996)	Data on six commercial finned tubes
748–2000	R-123	Sreepathi et al. (1996)	Effect of tin pitch and fin height
242–1366	R-12	Gogocin and Kabov (1996)	Data fat effect of fin radius
1024, 1654	R-11, R-12	Jung et al. (1999)	Circular fins and Turbo-C
1000	Steam	Liu et al. (1999)	Circular fin tube with wire mesh drainage strip
400	Steam	Das et al. (1999)	Effect of fin height for stainless steel tube
934	R-134a	Kumar et al. (2000)	Circular fins
390–1875	Steam, R-12, R-134A	Kumar et al. (2000)	Circular fins and spine fins
Wire–wrapped tube	Steam	Briggs et al. (2002, 2003)	1–6 mm wire pitch, 0.2, 0.4, 0.75, 1.0 wire diameter
390	Steam	Singh et al. (2001)	Four tubes in a vertical rank
390, 1560	Steam, R-134a	Kumar et al. (2002a)	Circular fins and spine fins; fins on total or only upper or lower half of lube

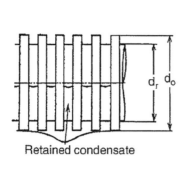 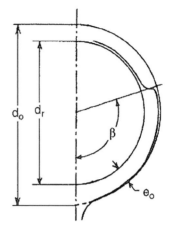

Fig. 4.3 Condensate flooding angle (β) on integral-finned tubes (from Webb and Kim 2005)

Table 4.3 Effect of drainage strip for methanol condensing on 2000 fins/m tube at $(T_s - T_w) = 5$ K (Webb and Kirn 2005)

Strip	H_d (mm)	C_b	h (W/m²-K)
None	–	0.62	6200
PVC	12.6	0.57	6900
Porous	12.6	0.32	10.200
Porous	4.0	0.42	7500

Figure 4.3 shows condensate flooding angle. Table 4.3 shows the effect of drainage strips. Figure 4.4 shows enhanced condensing tubes of Hitachi thermoexcel-C, Wieland GEWA-Sc, Wolverine Turbo-C and Sumitomo Trade-26D. Fig. 4.5 shows horizontal integral fin tubes for steam condensation by Jaber and Webb (1993). Figures 4.6 and 4.7 show R11 condensation coefficient of Webb et al. (1985), Sukhatme et al. (1990) and Webb and Murawski (1990). Wang et al. reported the performance of R113 in a unique finned tube having lateral ripples in the fins. They have observed that surface tension force pulls the condensate into the valleys of the ripples. Wildsmith (1980), Briggs and rose (1995, 1996), Briggs et al. (1995, 2002), Wen et al. (1994), Das et al. (1995), Renken and Raich (1996), Liu et al. (1999), Singh et al. (2001), Adamek and Webb (1990a), Abdullah et al. (1995), Rabas and Taborek (1999) and Kumar et al. (2000, 2002a, b) have worked on integral fin tubes for steam condensation.

Mehta and Rao (1979), Rao (1988), Withers and young (1971), Marto et al. (1979) and Yorkshire (1982) have worked with corrugated tubes for tube-side condensation enhancement. Condensation enhancement characteristics in corrugated titanium tubes having three different pitches have been studied by Kim (2019). He observed very low change in enhancement rate corresponding to high change in corrugation pitch. Kim (2019) also performed fouling test using silt obtained from power plants. He adopted continuous sponge cleaning method and observed nearly zero fouling resistance for a tube with 2.1 mm corrugation pitch.

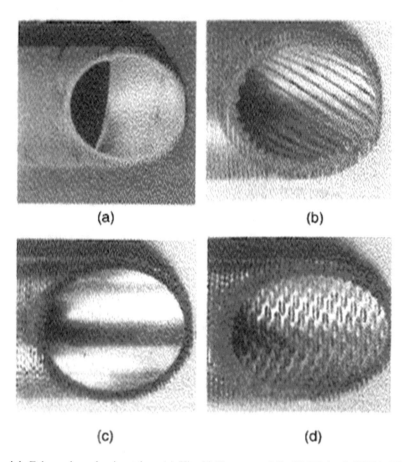

Fig. 4.4 Enhanced condensing tubes. (**a**) Hitachi Thermoexcel-C, (**b**) Wieland GEWA-SC, (**c**) Wolverine Turbo-C, (**d**) Sumitomo Tred-26D (from Webb and Kim 2005)

Figure 4.8 shows doubly enhanced tube geometries investigated by Marto et al. (1979). Surface tension drainage effect has been studied by Shah et al. (1999). Surface tension forces to enhance laminar flow condensation on a vertical surface were investigated by Gregorig (1954a, b). Figure 4.9 shows vertical fluted tube for enhancement. Webb (1979), Mori et al. (1979), Combs and Murphy (1978), Adamek and Webb (1990a), Thomas (1968), Canavos (1974) and Newson and Hodgson (1974) also have worked on vertical fluted tube and the enhanced laminar film condensation. Figures 4.10 and 4.11 show the results of the investigation with vertical fluted tube. Aly and Bedrose (1995) worked on the theory of enhanced condensation of water vapour on both vertical and horizontal spirally fluted tubes. For example the valley and crest regions of a sine wave flute were analysed. Surface tension and gravity were the driving forces for the condensation process. In the valley region of the flute, an adiabatic region was observed, and the condensation moved down the stream by gravity responsible for draining. The relation between

Fig. 4.5 Horizontal integral
fin tubes developed for
steam condensation by Jaber
and Webb (1993). (**a**)
Wolverine Korodense, (**b**)
copper Wieland 11-NW, (**c**)
copper-nickel Wieland
11-NW, (**d**) stainless steel
Wieland NW-16, (**e**) UOP
attached particle tube, (**f**)
Yorkshire MERT (multiple
enhanced roped tube)

condensation flow rate and film thickness was conjectured by solving Navier–Stokes
equations and the energy balance equation, for both crest and valley regions. They
observed as high as five times enhancement compared to the smooth tube case due to
fluting for the horizontal tubes. Their conclusion was substantiated by the available
experimental results of film condensation on horizontal finned tubes and twisted
vertical tubes; the configuration was akin to spirally fluted tubes. Their reported
results are found to be acceptable by good agreement of the theoretical predictions
with the experimental data.

Figure 4.12 shows the flute geometry and coordinate system used by Aly and
Bedrose (1995). Figure 4.13 shows the horizontal and vertical spirally fluted tubes
studied by Aly and Bedrose (1995). The works of Marto et al. (1979), Gregorig
(1954a, b), Hirasawa et al. (1980), Webb et al. (1982), Fathalah et al. (1987), Honda
et al. (1987), Rudy and Webb (1983, 1985), Adamek and Webb (1990a, b), Yau
et al. (1985) and Wanniarachchi et al. (1986) may be referred for further details of
enhanced film condensation of steam on spirally fluted tubes. Condensation of water
in horizontal tube bundles was investigated by Briggs et al. (2003) for heat transfer
and pressure drop characteristics, and the flow visualization has been carried out.
They observed that the heat transfer decreases and pressure drop increases with

Fig. 4.6 (a) R11
condensation coefficient on
standard copper integral fin
tubes (748, 1024 and
1378 fins/m) as reported by
Webb et al. (1985), (b) R11
condensation coefficient on
1417 fins/m copper integral
fin tubes with fin heights
between 0.46 and 1.22 mm
as reported by Sukhatme
et al. (1990)

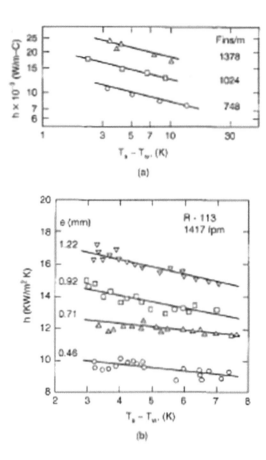

increase in condensation length while heat transfer and pressure drop both increased with decreasing tube diameter.

Axial wires on vertical smooth tubes have been used by Thomas (1967, 1968), Butizov et al. (1975), Rifert and Leontev (1976) and Kun and Ragi (1981). Wire wrap on horizontal tubes has been used by Thomas et al. (1979), Marto and Wanniarachchi (1984), Fujii et al. (1985, 1987), Marto et al. (1987), Briggs et al. (2002, 2003) and Rose (2002). Staub (1966) has studied surface tension effect in zero gravity. Briggs et al. (2003) studied condensation of steam on wire-wrapped tubes. A twofold enhancement has been observed. The effect of electric fields (EHD) has been studied by Yabe (1991). Sadek et al. (2011) observed an increase in enhancement factor of 2.7 in tube-side condensation of R134 using AC high voltage electric field. They reported that in high frequency range, the change in frequency has negligible impact on heat transfer while the heat transfer increased with frequency in low-frequency range. Detailed fundamentals of surface tension-drained condensation have been dealt with by Webb and Kim (2005) (book). Figure 4.14 shows finned plate with convex fin profiles, and this has been investigated by Gregorig (1954), Adamek (1981), Kedzierski (1987), Kedzierski and Webb

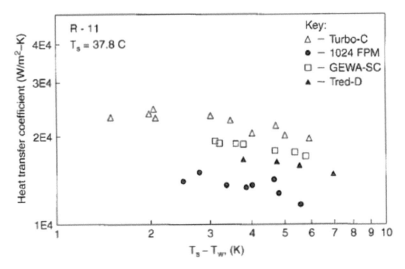

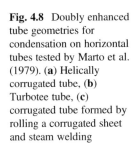

Fig. 4.7 R-11 condensation coefficient on copper enhanced horizontal tubes (from Webb and Murawski 1990)

Fig. 4.8 Doubly enhanced tube geometries for condensation on horizontal tubes tested by Marto et al. (1979). (**a**) Helically corrugated tube, (**b**) Turbotee tube, (**c**) corrugated tube formed by rolling a corrugated sheet and steam welding

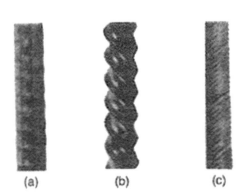

(1990), Webb et al. (1982, 1985), Adamek and Webb (1990a, b), Honda and Nozu (1987a), Zener and Lavi (1974) and Webb and Kedzierski (1990). Usually the fin profile shape is produced by electrostatic discharge machining (EDM). Table 4.4 shows the dimensions of commercially available fin tubes.

The fin profile shapes are not continuous functions, and complex models for the profiles have been developed. The models are Adamek and Webb model (Adamek and Webb 1990a, b), Adamek (1981), Honda and Nozu (1987a), Rose (1994), Briggs and Rose (1994), Cavallini et al. (1996), Belghazi et al. (2002), Srinivasan et al. (2002) and Panchal (1994). Figure 4.15 shows GEWA-C tube (Belghazi et al. 2002). The analytical models for horizontal integral fin tubes have been developed by Beatty and Katz (1948), Webb et al. (1982), Karkhu and Borovkov (1971), Rudy and Webb (1983, 1985), Webb et al. (1985), Honda and Nozu (1987a) and Adamek

Fig. 4.9 Vertical fluted
tubes. (**a**) Cross section of
fluted tube, (**b**) photo of
doubly fluted tube, (**c**) detail
of cross section (from Webb
and Kim 2005)

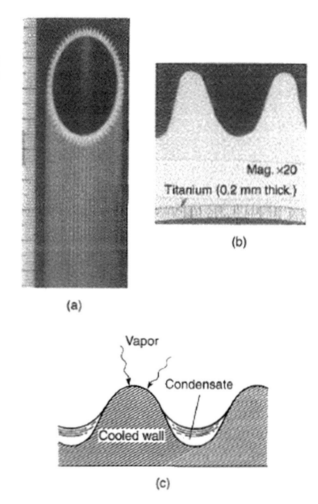

and Webb (1990b). The model of Beatty and Katz (1948) assumes that gravity force
drains the condensate from the fins and no condensate retention occurs on the lower
side of the tube. Precise surface tension-drained models have been developed by
Honda and Nozu (1987a) (Fig. 4.16) and Honda and Nozu (1987b). However, such
models have been reported by Jaber and Webb (1993, 1996) and Webb et al. (1985).
The comparison between theory and experiment has been done by Honda and Nozu
(1987a), Webb et al. (1985), Rudy and Webb (1985), Adamek and Webb (1990b),
Sukhatme et al. (1990) and Kumar et al. (2002a).

Fig. 4.10 (**a**) Vertical fluted
tube with drainage skirt; (**b**)
predicted and experimental
results on tube having
drainage skirt (from Mori
et al. 1979)

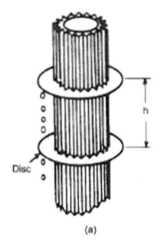

(a)

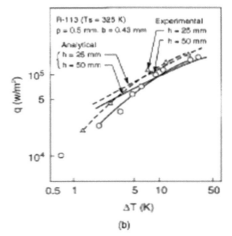

(b)

4.4   Horizontal Tube Banks

Webb (1984), Michael et al. (1989) and Browne and Bansal (1999) made a detailed
survey of condensation on tube banks which are affected by condensate inundation
and vapour velocity. Vapour shear causes considerable enhancement. Shekriladze
and Gomelauri (1966), Briggs et al. (1992), Lee and Rose (1984), Rose (1984),
Michael et al. (1989), McNaught and Cotchin (1989), Cavallini et al. (1993, 1994,
1995, 1996) and Honda et al. (1987, 1989, 1991, 1992) have developed correlation
for vapour shear effect based on theory. Condensate inundation may be without
shear (Kern 1958; Katz and Geist 1948; Brower 1985; Singh et al. 2001; Marto and
Wanniarachchi 1984; Kumar et al. 2002b; Webb and Murawski 1990; Honda and
Nozu 1987a; Honda et al. 2003; Belghazi et al. 2003). Condensate drainage pattern

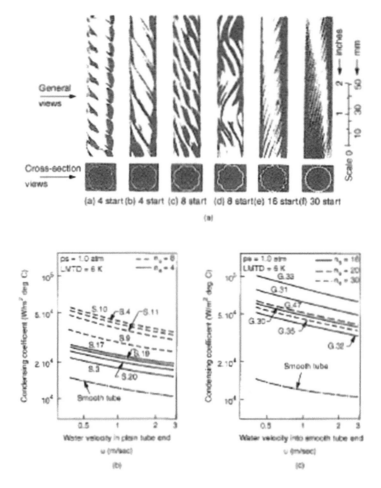

Fig. 4.11 (**a**) Tubes tested by Newson and Hodgson (1974); (**b**) condensing coefficient for 4 and 8 start tubes; (**c**) condensing coefficient for 16, 20 and 30 start tubes (from Newson and Hodgson 1974)

for a bank of integral fin tube has been studied by Honda and Nozu (1987a) and Honda et al. (2003) (Figs. 4.17 and 4.18). Further information on this may be obtained from Fujii (1991), Lee and Rose (1984), Michael et al. (1989), Honda and Nozu (1987a), Honda et al. (1991, 1992) and Roques and Thome (2003). After having the analytical model, the condensation coefficient may be predicted and much light has been shed in this direction by Honda and Nozu (1987a, b) and Adamek and Webb (1990b).

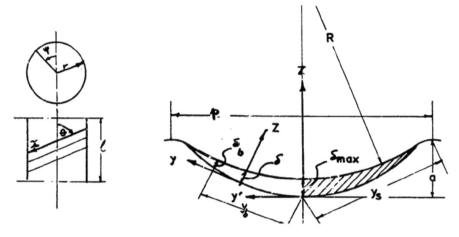

Fig. 4.12   Flute geometry and coordinate system (Aly and Bedrose 1995)

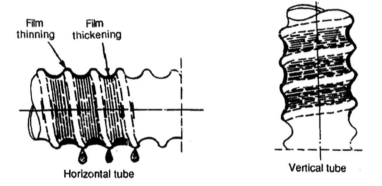

Fig. 4.13   Horizontal and vertical spirally fluted tubes (Aly and Bedrose 1995)

4.5   Conclusions

Surface tension forces influencing condensation has been discussed. Models for predicting integral horizontal finned tubes have been discussed. Experiments in tube bundles show low effect. Flow patterns have also been discussed. Characterization and prediction of condensation coefficient need further research.

Fig. 4.14 Illustration of finned plate with convex fin profiles designed for surface tension-drained condensation and rectangular condensate drainage channels (from Webb and Kim 2005)

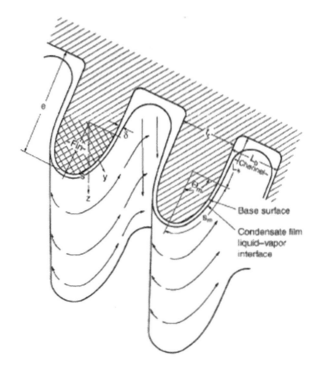

Table 4.4 Dimensions of commercially available internal fin tubes (Webb and Kim 2005)

Fins/metre (fins/m)	748	1024	1378
Outside diameter d_o (mm)	19.0	19.0	19.0
Area ratio $A_o/(\pi d_o L)$	2.91	3.60	3.18
Fin height e (mm)	1.53	1.53	0.89
Fin thickness at tip t_t (mm)	0.20	0.20	0.20
Fin thickness at base t_b (mm)	0.42	0.52	0.29
Aspect ratio, e/t_b	3.6	2.9	3.1

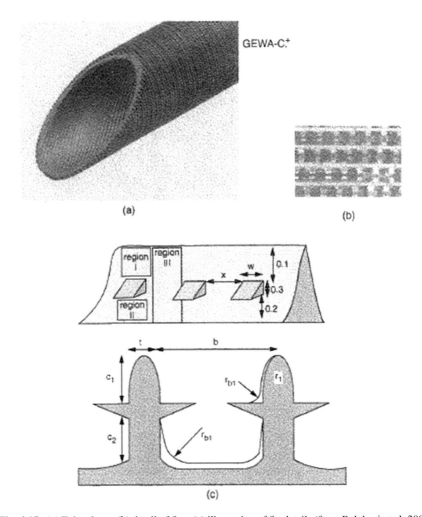

GEWA-C⁺

(a)

(b)

(c)

Fig. 4.15 (**a**) Tube photo, (**b**) detail of fins, (**c**) illustration of fin details (from Belghazi et al. 2002)

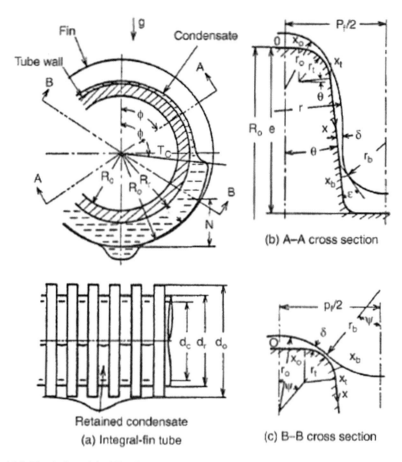

(b) A–A cross section

(a) Integral-fin tube

(c) B–B cross section

Fig. 4.16 Physical model of Honda and Nozu

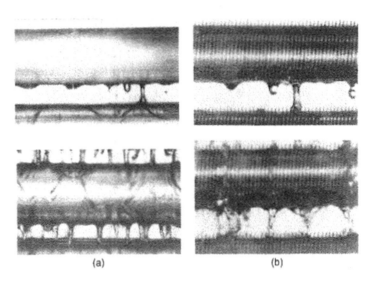

(a)   (b)

Fig. 4.17 Flow patterns observed by Honda and Nozu (1987a) for R-113 condensating at 0.12 MPa on 15.9 diameter plain and 1060 fins/m tubes. The top and bottom figures are for the first and 13th rows, respectively. (**a**) Plain tubes: first row ($Re = 19$), 13th row ($Re = 150$). (**b**) 1024 fins/m integral fin tubes: first row ($Re = 50$), 13th row ($Re = 550$) (from Honda and Nozu 1987b)

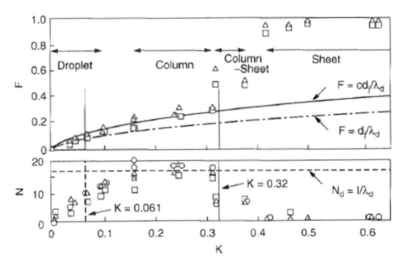

Fig. 4.18 Flow pattern map of parameters F and N vs. for R-113 condensing on 1060 fins/m tube (from Honda and Nozu 1987b)

References

Abdullah R, Cooper JR, Briggs A, Rose JW (1995) Condensation of steam and R113 on a bank of horizontal tubes in the presence of a noncondensing gas. Exp Thermal Fluid Sci 10(3):298–306

Adamek TA (1981) Bestimmung der Kondensationgrossen auf feingewellten Oberflachen zur Auslegun aptimaler Wandprofile. Wdnne Stoffbertrag 15:255–270

Adamek T, Webb RL (1990a) Prediction of film condensation on horizontal integral fin tubes. Int J Heat Mass Transf 33(8):1721–1735

Adamek T, Webb RL (1990b) Prediction of film condensation on vertical finned plates and tubes: a model for the drainage channel. Int J Heat Mass Transf 33(8):1737–1749

Aly NH, Bedrose SD (1995) Enhanced film condensation of steam on spirally fluted tubes. Desalination 101(3):295–301

Beatty KO, Katz DL (1948) Condensation of vapors on outside of finned tubes. Chem Eng Prog 44:55–70

Belghazi M, Bontemps A, Marvillet C (2002) Condensation heat transfer on enhanced surface tubes: experimental results and predictive theory. J Heat Transf 124(4):754–761

Belghazi M, Bontemps A, Marvillet C (2003) Experimental study and modelling of heat transfer during condensation of pure flnid and binary mixture on a bundle of horizontal finned tubes. Int J Refrig 26:214–223

Briggs A, Rose JW (1994) Effect of fin efficiency on a model for condensation heat transfer on a horizontal, integral-fin tube. Int J Heat Mass Transf 37:457–463

Briggs A, Rose JW (1995) Condensation performance of some commercial integral fin tubes with steam and CFC113. Exp Heat Transf Int J 8(2):131–143

Briggs A, Rose JW (1996) Condensation on low-fin tubes: effects of non-uniform wall temperature and interphase matter transfer. In: Process, enhancement, and multiphase heat transfer, pp 455–460

Briggs A, Wen XL, Rose JW (1992) Accurate heat transfer measurements for condensation on horizontal, integral-fin tubes. J Heat Transf 114(3):719–726

Briggs A, Huang XS, Rose JW (1995) An experimental investigation of condensation on integral-fin tubes: effect of fin thickness, height and thermal conductivity (No.CONF-950828). ASME, New York, NY

Briggs A, Wang HS, Rose J (2002) Film condensation of steam on a horizontal wire-wrapped tube. Heat Transf 4:123–128

Briggs A, Wang HS, Murase T, Rose JW (2003) Heat transfer measurements for condensation of steam on a horizontal wire-wrapped tube. J Enhanc Heat Transf 10(4)

Brower SK (1985) The effect of condensate inundation on steam condensation heat transfer in a tube bundle. Naval Postgraduate School, Monterey, CA

Brown CE, Matin SA (1971) The Effect of finite metal conductivity on the condensation heat transfer to falling water rivulets on vertical heat-transfer surfaces. J Heat Transf 93(1):69–76

Browne MW, Bansal PK (1999) An overview of condensation heat transfer on horizontal tube bundles. Appl Therm Eng 19(6):565–594

Butizov AI, Rifert VG, Leont'yev GG (1975) Heat transfer in steam condensation on wirefinned vertical surfaces. Heat Transf Soviet Res 7(5):116–120

Canavos VTC (1974) In: Afgan N, Schliinder EU (eds) Heat exchangers: design and theory sourcebook. McGraw-Hill, New York

Carnavos TC (1980) An experimental study: condensing R-11 on augmented tubes. ASME Paper No. 8o-HT-54

Carey VP (1992) Liquid-vapor phase-change phenomena. Hemisphere, Washington, DC

Cary JD, Mikic BB (1973) The influence of thermocapillary flow on heat transfer in film condensation. J Heat Transf 95(1):21–24

Cavallini A, Bella B, Longo GA, Rossetto L (1993) Pure vapour condensation of refrigerant 113 on a horizontal 2000 FPM integral finned tube. In: Taborek J, Rose, Tanasawa I (eds) Condensation and condenser design (Proceedings of the engineering foundation conference on condensation and condenser design). ASME, New York, pp 357–365

Cavallini A, Doretti L, Longo L (1994) Experimental heat transfer coefficients during external condensation of halogenated refrigerants on enhanced tubes. Heat transfer 1994, Proceedings of the 10th international heat transfer conference, vol 5, pp 7–12

Cavallini A, Bella B, Longo GA, Rossetto L (1995) Experimental heat transfer coefficients during condensation of halogenated refrigerants on enhanced tubes. J Enhanc Heat Transf 2:115–126

Cavallini A, Doretti L, Longo GA, Rossetto L (1996) A new model for forced-convection condensation on integral-fin tubes. J Heat Transf 118(3):689–693

Cheng B, Tao WQ (1994) Experimental study of R-152a film condensation on single horizontal smooth tube and enhanced tubes. J Heat Transf 116(1):266–270

Cheng WY, Wang CC Huang LW (1996) Film condensation of HCFC-22 on horizontal enhanced tubes, Int Comm Heat Mass Transf 23(1): pp. 79–90

Collier JG, Thome JR (1994) Convective boiling and condensation. Clarendon Press

Combs SK, Murphy RW (1978) Experimental studies of OTEC heat transfer condensation of ammonia on vertical fluted tubes. In: Proceedings of the fifth ocean thermal energy conversion (OTEC) conference, pp 20–22

Das AK, Meyer DW, Incheck GA, Marto PJ, Memory SB (1995) Effect of fin height and thermal conductivity on the performance of integral-fin tubes for steam condensation. In: Dhir VK (ed) Proceedings of the 30th national heat transfer conference, vol 308. ASME, New York, NY, pp 111–122

Das A, Incheck GA, Marto PJ (1999) The effect of fin height during steam condensation on a horizontal stainless steel integral-fin tube. J Enhanc Heat Transf 6(2–4)

Das A, Kilty HP, Marto PJ, Kumar A, Andeen GB (2000a) Dropwise condensation of steam on horizontal corrugated tubes using an organic self-assembled monolayer coating. J Enhanc Heat Transf 7(2)

Das AK, Kilty HP, Marto PJ, Andeen GB, Kumar A (2000b) The use of an organic self-assembled monolayer coating to promote dropwise condensation of steam on horizontal tubes. J Heat Transf 122(2):278–286

Dipprey DF, Sabersky RH (1963) Heat and momentum transfer in smooth and rough tubes at various Prandtl numbers. Int J Heat Mass Transf 6(5):329–353

Fathalah K, Aly SE, Darwish M, Radhwan A (1987) A theoretical study of enhanced condensation over horizontal fluted tubes. Desalination 65:25–42

Fujii T (1991) Representative physical properties for the condensate film and the vapor boundary layer. In: Theory of laminar film condensation. Springer, New York, NY, pp 153–172

Fujii T, Wang WC, Koyama S, Shimizu Y (1985) Heat transfer enhancement for gravity controlled condensation on a horizontal tube by a coiled wire. Trans J Soc Mech Eng Int J Ser B 51 (467):2436–2441

Glicksman LR, Mikic BB, Snow DF (1973) Augmentation of film condensation on the outside of horizontal tubes. AICHE J 19(3):636–637

Gogonin II, Kabov OA (1996) An experimental study of R-11 and R-12 film condensation on horizontal integral-fin tubes. J Enhanc Heat Transf 3:43–54

Gregorig R (1954a) Film condensation on finely rippled surfaces with consideration of surface tension. Z Angew Math Phys 5(1):36–49

Gregorig R (1954b) Hautkondensation an feingewellten Oberflächen bei Berücksichtigung der Oberflächenspannungen. Zeitschrift für angewandte Mathematik und Physik ZAMP 5(1):36–49

Griffith P (1985) Dropwise condensation. In: Rohsenow WM, Hartnett JP, Ganic EN (eds) Handbook of heat transfer fundamentals. McGraw-Hill, New York

Hanneman RJ (1977) Recent advances in dropwise condensation theory. ASME paper 77-WA/HT-21

Henrici K (1961) Kodensation von Frigen 12 und Frigen 22 an Glatten und Berippten Rohren. Dissertation, TU Karlsruhe

Hirasawa SH, Hijikata K, Mori Y, Nakayama W (1980) Effect of surface tension on condensate motion in laminar film condensation (study of liquid film in a small trough). Int J Heat Mass Transf 23:1471–1478

Honda H, Makishi O (1995) Effect of a circumferential rib on film condensation on a horizontal two-dimensional fin tube. J Enhanc Heat Transf 2(4)

Honda H, Nozu S (1985) A dimensionless correlation for film condensation on horizontal, low integral-fin tubes. Bull JSME 28:2824

Honda H, Nozu S (1987a) A prediction method for heat transfer during film condensation on horizontal low integral-fin tubes. J Heat Transf 109(1):218–225

Honda H, Nozu S (1987b) Effect of drainage strips on the condensation heat transfer performance of horizontal finned tubes. Heat Transfer Science and Technology, Hemisphere, New York, pp 455–462

Honda H, Takamatsu H, Kim K (1994) Condensation of CFC-11 and HCFC-123 in in-line bundles of horizontal finned tubes: effect of fin geometry. J Enhanc Heat Transf 1:197–210

Honda H, Nozu S, Mitsumori, K (1983). Augmentation of condensation on horizontal finned tubes by attaching a porous drainage plate. In: Proceedings of the ASME-JSME thermal engineering joint conference, vol 3, pp 289–296

Honda H, Uchima B, Nozu S (1987) A generalized prediction method for heat transfer during film condensation on a horizontal low-finned tube. In: Proceedings of American Society of Engineers and Japan Society of Mechanical Engineers, HI ASME-JSME Second Thermal Engineering Joint Conference, Honolulu, vol 4, pp 385–392

Honda H, Nozu S, Takeda Y (1989) A theoretical model of film condensation in a bundle of horizontal low finned tubes. J Heat Transf 11(2):525–532

Honda H, Uchima B, Nozu S, Nakata H, Torigoe E (1991) Film condensation of R-113 on in-line bundles of horizontal finned tubes. J Heat Transf 113(2):479–486

Honda H, Uchima B, Nozu S, Torigoe E, Imai S (1992) Film condensation of R-113 on staggered bundles of horizontal finned tubes. J Heat Transf 114(2):442–449

Honda H, Takata N, Takamatsu H, Kim JS, Usami K (2003) Effect of fin geometry on condensation of R407C in a staggered bundle of horizontal finned tubes. J Heat Transf 125(4):653–660

Iltscheff S (1971) Some experiments concerning the attainment of drop condensation with fluorinated refrigerants. Kältetechnik-Klimatisierung 23:237–241

Jaber MH, Webb RL (1993) Enhanced tubes for steam condensers. An Int J Exp Heat Transf 6(1):35–54

Jaber MH, Webb RL (1996) Steam condensation on horizontal integral-fin tubes of low thermal conductivity. J Enhanc Heat Transf 3(1)

Jung D, Kim C-B, Cho S, Song K (1999) Condensation heat transfer coefficients of enhanced tubes with alternative refrigerants for CFC11 and CFC12. Int J Refrig 22:548–557

Karkhu VA, Borovkov V (1971) Film condensation of vapor at finely-finned horizontal tubes. Heat Transf Soviet Res 3(2):183–191

Katz DL, Geist JM (1948) Condensation on six finned tubes in a vertical row. Trans ASME 70(8):907–914

Kedzierski MA (1987) Experimental measurements of condensation on vertical plates with enhanced fins. In: Boiling and condensation in heat transfer, Boston, MA

Kedzierski MA, Webb RL (1990) Practical fin shapes for surface-tension-drained condensation. J Heat Transf 112(2):479–485

Kern DQ (1958) Mathematical development of tube loading in horizontal condensers. AICHE J 4(2):157–160

Kim NH (2019) Steam condensation enhancement and fouling in titanium corrugated tubes. J Enhanc Heat Transf 26(1)

Kim KJ, Lefsaker AM, Razani A, Stone A (2001) The effective use of heat transfer additives for steam condensation. Appl Therm Eng 21(18):1863–1874

Kumar R, Varma HK, Mohanty B, Agrawal KN (2000) Condensation of R-134a vapor over single horizontal circular integral-fin tubes with trapezoidal fins. Heat Transfer Eng 21(2):29–39

Kumar R, Varma HK, Mohanty B, Agrawal KN (2002a) Augmentation of heat transfer during filmwise condensation of steam and R-134a over single horizontal finned tubes. Int J Heat Mass Transf 45(1):201–211

Kumar R, Varma HK, Mohanty B, Agrawal KN (2002b) Prediction of heat transfer coefficients during condensation of water and R-134a on single horizontal integral-fin tubes. Int J Refrig 25:111–126

Kun LC, Ragi EG (1981) US Patent No 4,253,519, US Patent and Trademark Office, Washington, DC

Lan Z, Ma X, Zhou XD, Wang M (2009) Theoretical study of dropwise condensation heat transfer: effect of the liquid-solid surface free energy difference. J Enhanc Heat Transf 16(1)

Lee WC, Rose JW (1984) Forced convection film condensation on a horizontal tube with and without non-condensing gases. Int J Heat Mass Transf 27(4):519–528

Liu X, Ma T, Zhang Z (1999) Investigation of enhancement of steam condensation heat transfer on finned tubes with porous drainage strips, Paper AJTE99-6350. In: Proceedings of the 5th ASMEIJSME joint thermal engineering conference, March 15–19, San Diego, CA

Lixin C, Jiehui Y (1998) A new treated surface for achieving dropwise condensation. J Enhanc Heat Transf 5(1)

Ma X, Chen J, Xu D, Lin J, Ren C, Long Z (2002) Influence of processing conditions of polymer film on dropwise condensation heat transfer. Int J Heat Mass Transf 45(16):3405–3411

Marto PJ, Wanniarachchi AS (1984) The use of wire-wrapped tubing to enhance steam condensation in tube bundles. In: Heat transfer in heat rejection systems, pp 9–16

Marto PJ, Reilly D, Fenner JH (1979) An experimental comparison of enhanced heat transfer condenser tubing. Adv Enhanc Heat Transf 16:1–9

Marto PJ, Mitrou E, Wanniarachchi AS, Katsuta M (1987) Film condensation of steam on a horizontal wire-wrapped tube. In: Proceedings of the 2nd ASME-JSME thermal engineering joint conference, vol 1, pp 509–516

Marto PJ, Zebrowski D, Wanniarachchi AS, Rose JW (1988) Film condensation of R-113 on horizontal finned tubes. In: Fundamentals of phase change: boiling and condensation, pp 583–592

Masuda H, Rose JW (1987) Condensation of ethylene glycol on horizontal integral-fin tubes. In: Proceedings of the 1987 ASME-JSME thermal engineering joint conference, vol 1. JSME and ASME, pp 525–530

McNaught JM, Cotchin CD (1989) Heat transfer and pressure drop in a shell and tube condenser with plain and low-fin tube bundles. Chem Eng Res Des (67):127–133

Medwell JO, Nicol AA (1965) Surface roughness effects on condensate films. ASME-AMER Soc Mech Eng 87(10):80

Mehta MH, Rao MR (1979) Heat transfer and frictional characteristics of spirally enhanced tubes for horizontal condensers. Adv Enhanc Heat Transf:11–21

Michael AG, Marto PJ, Wanniarachchi AS, Rose JW (1989) Effect of vapour velocity during condensation on horizontal smooth and finned tubes. In: Proceedings of the ASME winter annual meeting, San Francisco

Mori Y, Hijikata H, Hirasawa S, Nakayama W (1979) Optimized performance of condensers with outside condensing surface. In: Chenoweth JM et al (eds) Condensation heat transfer. ASME, New York, pp 55–62

Newson IH, Hodgson TD (1974) The development of enhanced heat transfer condenser tubing. Desalination 14(3):291–323

Nicol AA, Medwell JO (1966) The effect of surface roughness on condensing steam. The Canadian J Chem Engg 44(3):170–173

Notaro F (1979) US Patent No. 4,154,294, U.S. Patent and Trademark Office, Washington, DC

Nusselt W (1916) Die Oberflachenkondensation des Wasserdampfes. Zeitschr Ver Deut Ing 60:541–569

Panchal CB (1994) Generalized correlation for condensation on vertical fluted surfaces. Heat Transf Eng 15(4):19–23

Pearson JF, Withers JG (1969) New finned tube configuration improves refrigerant condensing. ASHRAE J 75:77–82

Rabas TJ, Taborek J (1999) Performance, fouling and cost considerations of enhanced tubes in in power-plant condensers. J Enhanc Heat Transf 6(2–4)

Rao MR (1988) Heat transfer and friction correlations for turbulent flow of water and viscous non-newtonian fluids in single–start spirally corrugated tubes. In: Proceedings of the national heat transfer conference, HTD-96, ASME, New York, vol 1, pp 677–683

Renken KJ, Aboye M (1993a) Experiments on film condensation promotion within thin inclined porous coatings. Int J Heat Mass Transf 36(5):1347–1355

Renken KJ, Aboye M (1993b) Experiments on film condensation promotion within thin inclined porous coatings. Int J Heat Mass Transf 14:48–53

Renken KJ, Mueller CD (1993) Measurements of enhanced film condensation utilizing a porous metallic coating. J Thermophys Heat Transf 7(1):148–152

Renken KJ, Raich MR (1996) Forced convection steam condensation experiments within thin porous coatings. Int J Heat Mass Transf 39(14):2937–2945

Rifert VG, Leontev G (1976) Analysis of heat transfer with steam condensing on a vertical surface with wires to promote heat transfer. Therm Eng 23(4):58–61

Roques JF, Thome JR (2003) Falling film transitions between droplet, column, and sheet flow modes on a vertical array of horizontal 19 FPI and 40 FPI low-finned tubes. Heat Transf Eng 24 (6):40–45

Rose JW (1984) Effect of pressure gradient in forced convection film condensation on a horizontal tube. Int J Heat Mass Transf 27(1):39–47

Rose JW (1994) An approximate equation for the vapour-side heat-transfer coefficient for condensation on low-finned tubes. Int J Heat Mass Transf 37(5):865–875

Rose JW (2002) An analysis of film condensation on a horizontal wire-wrapped tube. Chem Eng Res Des 80(3):290–294

Rudy TM, Webb RL (1983) Proceedings of the 1st ASME-JSME thermal engineering joint conference, vol 1, pp 373–377

Rudy TM, Webb RL (1985) An analytical model to predict condensate retention on horizontal integral-fin tubes. J Heat Transf 107(2):361–368

Sadek H, Cotton JS, Ching CY, Shoukri M (2011) In-tube convective condensation under AC high-voltage electric fields. J Enhanc Heat Transf 18(2)

Shah RK, Zhou SQ, Tagavi KA (1999) The role of surface tension in film condensation in extended surface passages. J Enhanc Heat Transf 6(2–4):179–216

Shekriladze IG, Gomelauri VI (1966) Theoretical study of laminar film condensation of flowing vapour. Int J Heat Mass Transf 9(6):581–591

Singh SK, Kumar R, Mohanty B (2001) Heat transfer during condensation of steam over a vertical grid of horizontal integral-fin copper tubes. Appl Therm Eng 21(7):717–730

Srinivasan PSS, Balasubramanian R, Gaitonde UN (2002) Correlation for laminar film condensation over single horizontal integral-fin copper tubes. Heat Transf 4:213–218

Sreepathi LK, Bapat SL, Sukhatme SP (1996) Heat transfer during film condensation of R-123 vapour on horizontal integral-fin tubes. J Enhanc Heat Transf 3(2):147

Staub PW (1966) Condensing heat transfer surface device, US Patent 3,289,752

Sukhatme SP, Jagadish BS, Prabhakaran P (1990) Film condensation of R-11 vapor on single horizontal enhanced condenser tubes. J Heat Transf 112(1):229–234

Takahashi A, Nosetani T, Miyata K (1979) Heat transfer performance of enhanced low finned tubes with spirally integral inside fins. Sumitomo Light Metal Tech Rep 20:59–65

Thomas DG (1967) Enhancement of film condensation heat transfer rates on vertical tubes by vertical wires. Ind Eng Chem Fundam 6(1):97–103

Thomas DG (1968) Enhancement of film condensation rate on vertical tubes by longitudinal fins. AICHE J 14(4):644–649

Thomas A, Lorenz JJ, Hillis DA, Young DT, Sather NP (1979) Performance tests of the 1 MWt shell and tube exchangers for OTEC. In: Proceedings of the 6th OTEC conference, Paper le

Utaka Y, Nishikawa T (2003) Measurement of condensate film thickness for solutal Marangoni condensation applying laser extinction method. J Enhanc Heat Transf 10(2)

Wang SP, Hijikata K, Deng SJ (1990) Experimental study on condensation heat transfer enhance-
    ment by Various Kinds of Integral Finned Tubes. In: Condensers and condensation, proceedings
    of the 2nd international symposium, pp xv–xxiii
Wang ZZ, Wei D, Hong F (2000) Experimental study of condensation heat transfer promotion on a
    fluted tube with thin porous coatings. Heat Transf Eng 21(4):46–52
Wanniarachchi AS, Marto PJ, Rose JW (1986) Film condensation of steam on horizontal finned
    tubes: effect of fin spacing. J Heat Transf 108(4):960–966
Webb RL (1979) A generalized procedure for the design and optimization of fluted Gregorig
    condensing surfaces. J Heat Transf 101:335–339
Webb RL (1984) The effects of vapor velocity and tube bundle geometry on condensation in shell-
    side refrigeration condensers. ASHRAE Trans 90(1B):39–59
Webb RL, Kim NY (2005) Principles of enhanced heat transfer. Taylor and Francis, New York
Webb RL, Kedzierski MA (1990) Practical fin shapes for surface tension drained condensation.
    J Heat Transf 112:479–485
Webb R, Murawski CG (1990) Row effect for R-11 condensation on enhanced tubes. J Heat Transf
    112(3):768–776
Webb RL, Keswani ST, Rudy TM (1982) Investigation of surface tension and gravity effects in film
    condensation. In: Proceedings of the 7th international heat transfer conference, Munich. Hemi-
    sphere, Washington, DC, pp 175–180
Webb RL, Rudy TM, Kedzierski MA (1985) Prediction of the condensation coefficient on
    horizontal integral-fin tubes. J Heat Transf 107(2):369–376
Wen XL, Briggs A, Rose JW (1994) Enhancement of condensation heat transfer on integral-fin
    tubes using radiused fin-root fillets. J Enhanc Heat Transf 1(2)
Wildsmith G (1980) Open Discussion section. In: Marro PJ, Nunn RH (eds) Power condenser heat
    transfer technology. Hemisphere, New York, pp 463–468
Withers JG, Young EH (1971) Steam condensing on vertical rows of horizontal corrugated and
    plain tubes. Application in desalination of water. Indust Eng Chem Process Des Dev 10
    (1):19–30
Yabe A (1991) Active heat transfer enhancement by applying electric fields. In: Proceedings of the
    1991 ASME JSME thermal engineering joint conference
Yan J, Wang J, Hu S, Chong D, Liu J (2011) Marangoni condensation heat transfer of water-ethanol
    mixture vapor. J Enhanc Heat Transf 18(4)
Yau KK et al (1985) J Heat Transf:108–377
Yau KK, Cooper JR, Rose JW (1986) Horizontal plain and low-finned condenser tubes—effect of
    fin spacing and drainage strips on heat transfer and condensate retention. J Heat Transf 108
    (4):946–950
Yorkshire (1982) YIM heat exchanger tubes: design data for horizontal rope tubes in steam
    condensers. Technical Memorandum 3, Yorkshire Imperial Metals, Ltd., Leeds, England
Zener C, Lavi A (1974) Drainage systems for condensation. J Heat Transf 96:209–205
Zhao Q, Burnside BM (1994) Dropwise condensation of steam on ion implanted condenser
    surfaces. Heat Recov Syst CHP 14(5):525–534

Chapter 5
Convective Condensation

For condensation in a tube, convective condensation rather than vapour space condensation is to be considered, and the convective effects are associated with the influence of shear stress on the liquid–vapour interface. With the gravity force stronger than viscous shear forces, the liquid phase tends to stratify in horizontal tube. Shear force effect on the vapour is the strongest near the inlet of the tube, the vapour velocity being the highest. Figure 5.1 shows flow patterns for condensation in a horizontal tube. Inside tubes having internally finned geometry has been studied by Vrable et al. (1974), Reisbig (1974), Royal and Bergles (1978a, b), Luu and Bergles (1979), Akers et al. (1959), Said and Azer (1983) and Kaushik and Azer (1988, 1989, 1990). Further investigations have been done by Sur and Azer (1991), Kimura (1979), Smit and Meyer (2002) and Shinohara and Tobe (1985). Tables 5.1 and 5.2 give the details of some investigations made with internally finned geometry, and Fig. 5.2 gives some test results of Royal and Bergles (1978a).

Wire loop finned annulus has been studied by Honda et al. (1988) and Nozu et al. (1995). Twisted tape insert has been used by Lopina and Bergles (1969). They have observed enhancement in condensation. Roughness tests have been made by Luu and Bergles (1979), Fenner and Ragi (1979), Shinohara and Tobe (1985) and Hinton et al. (1995). Wire coil inserts, coiled tubes and return bends have been used by Wang (1987), Thomas (1967), Agrawal et al. (1998), Brdlik and Kakabaev (1964), Miropolskii and Kurbanmukhamedov (1975) and Traviss and Rohsenow (1971). Some of them have given predictive correlations.

Yasuda et al. (1990), Shinohara and Tobe (1985), Khanpara et al. (1987), Schlager et al. (1988, 1989a, b, 1990a, b), Eckels and Pate (1991), Cavallini et al. (2000, 2002, 2003), Newell and Shah (2001), Webb (1999) and Liebenberg et al. (2000) have studied microfin tube applications for convective condensation. Table 5.3 gives optimum microfin configuration. Yang et al. (2003) investigated condensation heat transfer in plain plate surface and surface with microfins having fin tip radius for different positions of the heated surface (vertical, downward and upward). They observed very slight difference for downward and upward positions for all plates as they were immersed in thick condensate film while due to surface

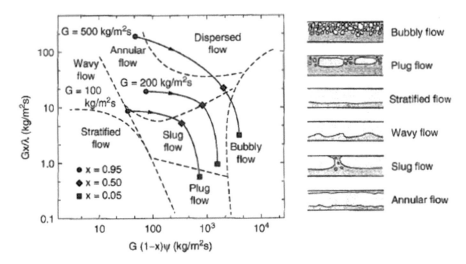

Fig. 5.1 Flow patterns for condensation in a horizontal tube drawn on the Baker (1954) flow pattern map (from Carey 1992)

Table 5.1 Geometries tested by Royal and Bergles (1978a)

Code	d_i (mm)	Geometry	e (mm)	n	α (deg)	A/A_p
A	15.9	Smooth	None	None	0	1
B	15.9	Twisted tape	None	None	18.7	1
C	15.9	Twisted tape	None	None	9.3	1
D	15.9	Internal fin	0.60	32	2.95	1.70
E	12.8	Internal fin	1.74	6	5.25	1.44
F	12.8	Internal fin	1.63	6	0	1.44
G	15.9	Internal fin	1.45	16	3.22	1.73

Table 5.2 Internal fin geometries tested by Said and Azer (1983)

Code	d_i (mm)	Fin height (mm)	No. of fins	Helix angle (deg)	A/A_p
2	13.84	1.58	10	0	1.50
3	17.15	1.80	16	9.7	1.89
4	19.87	1.98	16	12.3	1.76
5	25.38	2.13	16	19.4	1.64

tension brings down the condensate to the valley region of the fins in case of vertical plate and helps in heat transfer augmentation. Tsuchida et al. (1993), Houfuku et al. (2001), Chamra et al. (1996a, b) and Ishikawa et al. (2002) have investigated for optimization of internal geometry convective condensation mechanism in microfin tubes has been studied by Moser et al. (1998), Friedel (1979), Webb et al. (1998), Webb (1999), Bhatia and Webb (2001) and Zivi (1964). Special microfin geometries

Fig. 5.2 Test results of
Royal and Bergles (1978a)
for steam condensation
(4 bar) in horizontal tubes
containing internal fins and
twisted tapes. (**a**) Heat
transfer coefficient based on
plain tube area and (**b**)
pressure drop

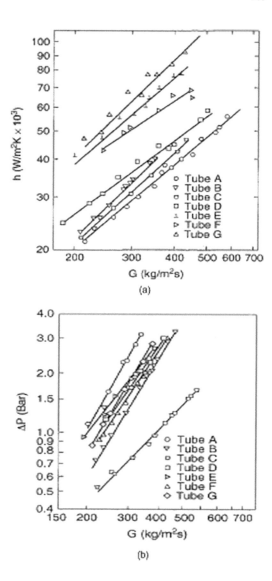

have been used by Chamra et al. (1996a, b), Ferreira et al. (2003), Tang et al. (2000a), Xin et al. (1998), Muzzio et al. (1998), Ebisu (1999), Miyara et al. (2003), Chiang (1993) and Graham et al. (1999).

Flat plate automotive condenser is very important, and these have been studied by Ohara (1983) and Hoshino et al. (1991). Figure 5.3 shows aluminium tubes used in automotive and extruded tube which have been obtained by Yang and Webb (1996, 1997), Webb and Yang (1995), Kim et al. (2000, 2003) and Webb and Ermis (2001). Figure 5.4 shows refrigerant circuiting method. Koyama et al. (2003) and Wang et al. (2002) obtained R-134a condensation in extruded aluminium tube. Their data

Table 5.3 Optimum microfin configuration (Webb and Kim 2005)

References	Geometry	Test conditions	Optimum configuration (vaporization)	Optimum configuration (condensation)
Ito and Kimura (1979)	$d_o = 12.7$	R-22	$\alpha = 7°$	
	$d_i = 11.2$	$40 < G < 203$	$p_f = 0.5\text{--}1.0$	
			$e = 0.2$	
Yasuda et al. (1990)	$d_o = 9.52,$ 7.94	R-22	$\alpha = 18°$	$\alpha = 30°$
	$d_i = 8.8,$ 7.22	$100 < G < 300$	$d_o = 9.52$	$d_o = 9.52$
	$\alpha = 40°$	$q = 10 \text{ kW/m}^2$	$n_f = 60$	$n_f = 50$
			$p_f = 0.44$	$p_f = 0.48$
Morita et al. (1993)	$d_o = 9.52,$ 7.00, 4.0	R-22	$d_o = 7.0$	$d_o = 9.52$
	$d_i = 8.92$ 6.4, 3.4		$n_f = 50$	$n_f = 50$
	$\alpha = 18°$		$d_o = 4.0$	$d_o = 4.0$
			$n_f = 34$	$n_f = 34$
Tsuchida et al. (1993)	$d_o = 9.52$	R-22	$\alpha = 7°$	$\alpha = 25°$
	$(d_i = 8.92)$	$q = 10 \text{ kW/m}^2$	$e = 0.25$	$e = 0.25$
	$\beta = 40°$	$150 < G < 600$	$n_f = 70 \ (p_f = 0.40)$	$n_f = 80 \ (p_f = 0.32)$
	$d_e = 7.0$		$\alpha = 18°$	$\alpha = 25°$
	$(d_i = 6.4)$		$e = 0.2$	$e = 0.18$
	$\beta = 40°$		$n_f = 60 \ (p_f = 0.32)$	$n_f = 60 \ (p_f = 0.30)$
	$d_o = 5.0$		$\alpha = 6°$	$\alpha = 10°$
	$(d_i = 4.2)$		$e = 0.15$	$e = 0.15$
	$\beta = 40°$		$n_f = 45 \ (p_f = 0.29)$	$n_f = 45 \ (p_f = 0.29)$
	$d_o = 4.0$		$\alpha = 6°$	$\alpha = 10°$
	$(d_i = 3.3)$		$e = 0.13$	$e = 0.13$
	$\beta = 40°$		$n_f = 40 \ (p_f = 0.26)$	$n_f = 40 \ (p_f = 0.26)$
Chamra et al. (1996a, b)	$d_o = 15.88$	R-22	$\alpha = 20°$	$\alpha = 27°$
	$e = 0.35$	$72 < G < 289$		
	$p = 0.58$			
	$\alpha = 30°$			
Houfuku et al. (2001)	$d_o = 7$	R-410A	$\alpha = 16°, \beta = 22°$	$\alpha = 16°, \beta = 22°$
	$(d_t = 6.4)$	$G = 250$	$n_f = 54 \ (p_f = 0.36)$	$n_f = 54 \ (p_f = 0.36)$
			$e = 0.22$	$e = 0.22$
Ishikawa et al. (2002)	$d_o = 7$	R-22	$n_f = 80$	$n_f = 80$
	$(d_i = 6.4)$	$160 < G < 320$	$(p_f = 0.24)$	$(p_f = 0.24)$
	$\alpha = 16°,$ $\beta = 10°$			
	$e = 0.24$			

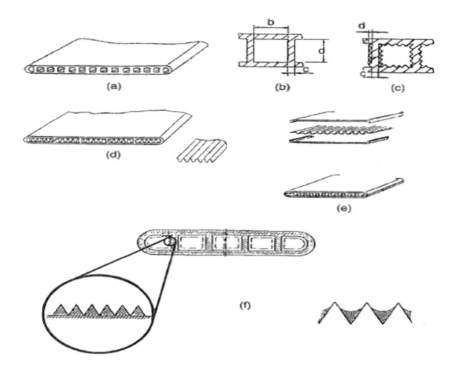

Fig. 5.3 Aluminium tube used in automotive, refrigerant condensers: (**a**) extruded tube with partitions, (**b**) details of part a tube, (**c**) details of tube having microgrooves, (**d**) extruded tube with corrugated insert, (**e**) tube made by brazing a corrugated strip in a tube formed from flat strip and seam welded, (**f**) illustration of surface tension-drained condensation on the tips of the microgrooves (from Webb and Kim 2005)

do not tally with the results predicted by Haraguchi et al. (1994). Jaster and Kosky (1976) correlation predicts the data in stratified region. Several predictive models of condensation in flat tubes have been given and discussed by Young and Webb (1996, 1997), Webb and Young (1995), Kim et al. (2000, 2003), Koyama et al. (2003), Webb (1999), Zhang (1998), Akers et al. (1959), Moser et al. (1998), Webb et al. (1998), Zhang and Webb (2001), Friedel (1979), Haraguchi et al. (1994), Mishima and Hibiki (1995), Wang et al. (2002), Traviss et al. (1973) and Chato (1962) and Zivi (1964). A model of U-shaped minichannel has been studied by Shao and Zhang (2011) for evaporation and condensation heat transfer of thermally induced oscillatory flow.

Figure 5.5 shows plate heat exchanger developed for condensation. Non-condensable gases impose additional thermal resistance. However, mixing in the gas film substantially reduces the thermal resistance. Chang and Spencer (1971) have studied the effect of non-condensable gases during convective condensation. Munoz-Cobo et al. (2004) presented a new model to evaluate steam condensation in finned tubes. The presence of non-condensable gases and aerosols has been

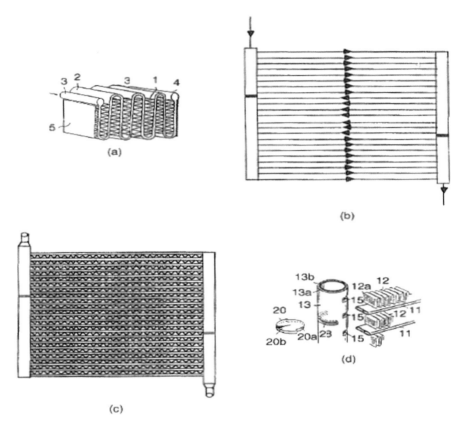

Fig. 5.4 (**a**) Serpentine refrigerant circuiting method, (**b**) parallel flow method of circuiting, (**c**) condenser having three refrigerant circuits, (**d**) separator plate brazed into header for multipass circuiting (from Webb and Kim 2005)

considered. Jokar et al. (2004) presented condensation heat transfer and pressure drop characteristics for brazed plate heat exchangers. Predictive equations for circular tubes have been developed by Schlager et al. (1988). Correlation for high internal fins has been developed by Vrable et al. (1974), Royal and Bergles (1978a), Akers et al. (1959), Luu and Bergles (1980), Boyko and Kruzhilin (1967), Kaushik and Azer (1988), Said and Azer (1983) and Venkatesh (1984). Honda et al. (1988), Kaushik and Azer (1989), Royal and Bergles (1978a), Said and Azer (1983) and Sur and Azer (1991) have developed correlations for wire loop internal fins. Correlation for twisted tapes during convective condensation has been developed by Royal and Bergles (1978a), Akers et al. (1959), Luu and Bergles (1980), Boyko and Kruzhilin (1967) and Said and Azer (1983); all of them have worked mostly with different refrigerants.

Helical rib roughness correlation has been given by Luu and Bergles (1980) using single-phase roughness correlation of Webb et al. (1971). Existing correlation for predicting the convective condensation coefficient in microfin tubes have been

Fig. 5.5 Plate heat exchanger developed for condensation against liquid coolant by Uehara and Sumitomo (1985). (**a**) End view showing plate configuration with condensing vapour (A) and coolant (B) steams. (**b**) Measured overall heat transfer coefficient for condensation of ammonia and R-114

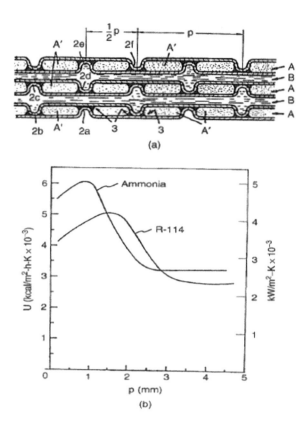

discussed extensively by Cavallini et al. (2002). The first ever correlation for microfin tubes was given by Cavallini et al. (1996). This is an extension of plain tube correlation given by Cavallini and Zecchin (1974). Cavallini et al. (1999) developed a model to predict pressure drop in microfin tube as an extension of plain tube correlation of Friedel (1979) and Sardesai et al. (1982). In this connection, Colebrook et al. (1939) may be used to calculate the Moody friction factor. Tang et al. (2017) observed increased flow condensation heat transfer rates in microfin tubes over those of smooth tube for refrigerants R22, R134a and R410A. They have also reported that the effect of refrigerant on heat transfer augmentation was insignificant. Wu et al. (2004) discussed the condensation heat transfer enhancement characteristics in microfin tubes using R22 as working fluid. Samokhvalov et al. (2018) reviewed different numerical models on condensation heat transfer characteristics of hydrocarbon mixture in horizontal, inclined and spiral-wound tubes. Brognaux et al. (1997), Chamra et al. (1996a), Webb (1999), Moser et al. (1998), Zivi (1964), Tang et al. (2000b), Nozu and Honda (2000), Honda and Nozu (1987), Wang et al. (2002), Wang and Honda (2003), Yu and Koyama (1998), Goto et al. (2003) and Miyara et al. (2000) are the additional references for microfins used in convective condensation. Lopez-Belchi et al. (2015) studied the effect of non-uniform condensation of refrigerant in multiport minichannel tubes. They reported that non-uniform condensation of refrigerant R134a resulted in reduced heat transfer rates.

Closure Internal enhancements of convective condensation by using cost-effective microfin tube give the highest heat transfer performance with the lowest accompanying pressure drop. Predictive models have been developed using surface geometry and fluid properties parameter. Optimization has also been done. Twisted tapes do not perform as effectively for convective condensation as the performance obtained from microfin tubes. Semi-empirical correlations perform rather well in case of twisted tapes.

References

Agrawal KN, Kumar A, Behabadi MA, Varma HK (1998) Heat transfer augmentation by coiled wire inserts during forced convection condensation of R-22 inside horizontal tubes. Int J Multiphase Flow 24(4):635–650

Akers WW, Deans HA, Crosser OK (1959) Condensing Heat Transfer Within Horizontal Tubes. Chem Eng Prog Symp Ser Heat Transfer 55(29):171

Baker O (1954) Simultaneous flow of oil and gas. Oil Gas J 53:185–195

Bhatia RS, Webb RL (2001) Numerical study of turbulent flow and heat transfer in micro-fin tubes—Part 1, model validation. J Enhanc Heat Transf 8(5)

Boyko LD, Kruzhilin GN (1967) Heat transfer and hydraulic resistance during condensation of steam in a horizontal tube and in a bundle of tubes. Int J Heat Mass Transf 10(3):361–373

Brdlik PM, Kakabaev A (1964) An experimental investigation of the condensation of steam in coils. Int Chem Eng 4(2):236–239

Brognaux LJ, Webb RL, Chamra LM, Chung BY (1997) Single-phase heat transfer in micro-fin tubes. Int J Heat Mass Transf 40(18):4345–4357

Carey VP (1992) Liquid-vapor phase-change phenomena. Hemisphere Pub. Co., Washington, DC

Cavallini A, Zecchin R (1974) A dimensionless correlation for heat transfer in forced convection condensation. In: Proceedings of the sixth international heat transfer conference, vol 3, pp 309–313

Cavallini A, Doretti L, Longo GA, Rossetto L (1996) A new model for forced-convection condensation on integral-fin tubes. J Heat Transf 118(3):689–693

Cavallini A, Del Col D, Doretti L, Longo GA, Rossetto L (1999) A new computational procedure for heat transfer and pressure drop during refrigerant condensation inside enhanced tubes. J Enhanc Heat Transf 6(6)

Cavallini A, Del Col D, Longo GA, Rossetto L (2000) Heat transfer and pressure drop during condensation of refrigerants inside horizontal enhanced tubes. Int J Refrig 23:4–25

Cavallini A, Censi G, Del Col D, Doretti L, Longo AG, Rossetto L (2002) Condensation heat transfer and pressure drop inside channels for AC/HP application. In: Heat transfer Proceedings of the 12th international heat transfer conference, vol 1, pp 171–186

Cavallini A, Censi G, Del Col D, Doretti L, Longo GA, Rossetto L, Zilio C (2003) Condensation inside and outside smooth and enhanced tubes, a review of recent research. Int J Refrig 26 (4):373–392

Chamra LM, Webb RL, Randlett MR (1996a) Advanced micro-fin tubes for condensation. Int J Heat Mass Transf 39(9):1839–1846

Chamra LM, Webb RL, Randlett MR (1996b) Advanced micro-fin tubes for evaporation. Int J Heat Mass Transf 39(9):1827–1838

Chang KI, Spencer DL (1971) Effect of regularly spaced surface ridges on film condensation heat transfer coefficients for condensation in the presence of noncondensable gas. Int J Heat Mass Transf 14(3):502–505

Chato JC (1962) Laminar condensation inside horizontal and inclined tubes. ASHRAE J 52:60

Chiang R (1993) Heat transfer and pressure drop during evaporation and condensation of refrigerant-22 in 7.5 mm and 10 mm diameter axial and helical grooved tubes. AIChE Symp Ser 89(259):205–205

Colebrook CF, Blench T, Chatley H, Essex EH, Finniecome JR, Lacey G, Williamson J, Macdonald GG (1939) Turbulent flow in pipes, with particular reference to the transition region between the smooth and rough pipe laws (includes plates). J Inst of Civ Eng 12(8):393–422

Ebisu T (1999) Evaporation and condensation heat transfer enhancement for alternative refrigerants used in air-conditioning machines. In: Heat transfer enhancement of heat exchangers. Springer, Dordrecht, pp 579–600

Eckels SJ, Pate MB (1991) In-tube evaporation and condensation of refrigerant-lubricant mixtures of HFC-134a and CFC-12. ASHRAE Trans 97(2):62–70

Fenner GW, Ragi EG (1979) Enhanced tube inner surface heat transfer device and method. US Patent 4,154,293

Ferreira CI, Newell TA, Chato JC, Nan X (2003) R404A condensing under forced flow conditions inside smooth, microfin and cross-hatched horizontal tubes. Int J Refrig 26(4):433–441

Friedel L (1979) Improved friction pressure drop correlation for horizontal and vertical two-phase pipe flow. In: Proceedings of the European two-phase flow group meet, Ispra, Italy

Goto M, Inoue N, Yonemoto R (2003) Condensation heat transfer of R410A inside internally grooved horizontal tubes. Int J Refrig 26(4):410–416

Graham D, Chato JC, Newell TA (1999) Heat transfer and pressure drop during condensation of refrigerant 134a in an axially grooved tube. Int J Heat Mass Transf 42(11):1935–1944

Haraguchi H, Koyama S, Fujii T (1994) Condensation of refrigerants HCFC22, HFC134a and HCFC123 in a horizontal smooth tube. 2nd Report. Proposals of empirical expressions for local heat transfer coefficient. Trans JSME (B) 60(574):245–225

Hinton DL, Conklin JC, Vineyard EA (1995) Condensation of refrigerants flowing inside smooth and corrugated tubes. Proc ASME/JSME Therm Eng Joint Conf 2:439–446

Honda H, Nozu S (1987) A prediction method for heat transfer during film condensation on horizontal low integral-fin tubes. J Heat Transf 109(1):218–225

Honda H, Nozu S, Matsuoka Y, Aomi T (1988) Condensation of refrigerants R-11 and R-113 in horizontal annuli with an enhanced inner tube. In: Shah RK, Ganie EN, Yang KT (eds) Proc. 1st world Conf. on experimental heat transfer: fluid mechanics and thermodynamics. Elsevier Science, New York, pp 1069–1076

Hoshino R, Sasaki H, Yasutake K (1991) Condenser for Use in Car Cooling System. U.S. Patent 5,025,855, assigned to Showa Aluminum Co, Japan

Houfuku M, Suzuki Y, Inui K (2001) High performance, light weight thermo-fin tubes for air-conditioners using alternative refrigerants. Hitachi Cable Rev 20:97–100

Ishikawa S, Nagahara K, Sukumoda S (2002) Heat transfer and pressure drop during evaporation and condensation of HCFC22 in horizontal copper tubes with many inner fins. J Enhanc Heat Transf 9(1):17–24

Ito M, Kimura H (1979) Boiling heat transfer and pressure drop in internal spiral-grooved tubes. Bull JSME 22(171):1251–1257

Jaster H, Kosky PG (1976) Condensation heat transfer in a mixed flow regime. Int J Heat Mass Transf 19(1):95–99

Jokar A, Eckels SJ, Honsi MH, Gielda TP (2004) Condensation heat transfer and pressure drop of brazed plate heat exchangers using refrigerant R-134a. J Enhanc Heat Transf 11(2)

Kaushik N, Azer NZ (1988) A general Heat Transfer correlation for condensation inside internally finned tubes. ASHRAE Trans 94(2):261–279

Kaushik N, Azer NZ (1989) An analytical heat transfer prediction model for condensation inside longitudinally finned tubes. ASHRAE Trans 95(2):516–523

Kaushik N, Azer NZ (1990) A general pressure drop correlation for condensation inside internally finned tubes. ASHRAE Trans 96(1):242–255

Khanpara JC, Pate MB, Bergles AE (1987) A comparison of local evaporation heat transfer enhancement for a micro-fin tube using refrigerants 22 and 113. In: Boiling and condensation in heat transfer equipment, ASME Papers, pp 31–39

Kim NH, Cho JP, Kim JO (2000) R-22 condensation in flat aluminium multi-channel tubes. J Enhanc Heat Transf 7(6):427–438

Kim NH, Cho JP, Kim JO, Youn B (2003) Condensation heat transfer of R-22 and R-410A in flat aluminum multi-channel tubes with or without micro-fins. Int J Refrig 26(7):830–839

Koyama S, Kuwahara K, Nakashita K, Yamamoto K (2003) An experimental study on condensation of refrigerant R134a in a multi-port extruded tube. Int J Refrig 26(4):425–432

Liebenberg L, Bergles AE, Meyer JP (2000) A review of refrigerant condensation in horizontal micro-fin tubes. In: Garimella S, Von Spakovsky M, Somasundaram S (eds) Proceedings of the ASME advanced energy systems division, AES, vol 40, pp 155–168

Lopez-Belchi A, Vera-Garcia F, Garcia-Cascales JR (2015) Non-uniform condensation of refrigerant R134a in mini-channel multiport tubes: two-phase pressure drop and heat transfer coefficient. J Enhanc Heat Transf 22(5)

Lopina RF, Bergles AE (1969) Heat transfer and pressure drop in tape-generated swirl flow of single-phase water. J Heat Transf 91(3):434–442

Luu M, Bergles AE (1979) Experimental study of the augmentation of in-tube condensation of R-113. ASHRAE Trans 85(2):132–146

Luu M, Bergles AE (1980) Enhancement of horizontal in-tube condensation of R-113. ASHRAE Trans 86:293–312

Miropolskii ZL, Kurbanmukhamedov A (1975) Heat transfer with condensation of steam within coils. Therm Eng 5:111–114

Mishima K, Hibiki T (1995) Effect of inner diameter on some characteristics of air-water two-phase flows in capillary tubes. Nippon Kikai Gakkai Ronbunshu B Hen 61(589):3197–3204

Miyara A, Nonaka K, Taniguchi M (2000) Condensation heat transfer and flow pattern inside a herringbone-type micro-fin tube. Int J Refrig 23(2):141–152

Miyara A, Otsubo Y, Ohtsuka S, Mizuta Y (2003) Effects of fin shape on condensation in herringbone micro-fin tubes. Int J Refrig 26(4):417–424

Morita H, Kito Y, Satoh Y (1993) Recent improvements in small bore inner grooved Coppe tube. Tube and Pipe Technology, pp 53–57

Moser KW, Webb RL, Na B (1998) A new equivalent Reynolds number model for condensation in smooth tubes. J Heat Transf 120(2):410–417

Munoz-Cobo JL, Escriva A, Herranz LE (2004) Mechanistic modeling of steam condensation onto finned tube heat exchangers in presence of noncondensable gases and aerosols, under cross-flow conditions: aerosol fouling and noncondensable gases effects on heat transfer. J Enhanc Heat Transf 11(1)

Muzzio A, Niro A, Arosio S (1998) Heat transfer and pressure drop during evaporation and condensation of R22 inside 9.52-mm OD micro-fin tubes of different geometries. J Enhanc Heat Transf 5(1):39–52

Newell TA, Shah RK (2001) An assessment of refrigerant heat transfer, pressure drop and void fraction effects in micro-fin tubes. HVAC&R Res 7(2):125–154

Nozu S, Honda H (2000) Condensation of refrigerants in horizontal, spirally grooved microfin tubes: numerical analysis of heat transfer in the annular flow regime. J Heat Transf 122(1):80–91

Nozu S, Honda H, Nishida S (1995) Condensation of a zeotropic CFC 114-CFC 113 refrigerant mixture in the annulus of a double-tube coil with an enhanced inner tube. Exp Thermal Fluid Sci 11(4):364–371

Ohara 1983. Heat exchanger. Japanese Patent 58,221,390, assigned to Nippondenso Co

Reisbig RL (1974) Condensing heat transfer augmentation inside splined tubes. In: AIAA/ASME Thermophysics and Heat Transfer Conference, paper 74-HT-7

Royal JH, Bergles AE (1978a) Augmentation of horizontal in-tube condensation by means of twisted-tape inserts and internally finned tubes. J Heat Transf 100(1):17–24

Said SA, Azer NZ (1983) Heat transfer and pressure drop during condensation inside horizontal tubes with twisted tape inserts. ASHRAE Trans 89(1):96–113

Samokhvalov Y, Kolesnikov A, Krotov A, Parkin A, Navasardyan ES, Arkharov IA (2018) Heat transfer in the structure of a spiral-wound heat exchanger for liquefied natural gas production: review of numerical models for the heat-transfer coefficient of condensation for a hydrocarbon mixture in a horizontal tube. J Enhanc Heat Transf 25(2)

Sardesai RG, Owen RG, Pulling DJ (1982) Pressure drop for condensation of a pure vapour in downflow in a vertical tube. UKAEA Atomic Energy Research Establishment

Schlager LM, Pate MB, Bergles AE (1988) Evaporation and condensation of refrigerant-oil mixtures in a smooth tube and a micro-fin tube. ASHRAE Trans 94(Part 1):149–166

Schlager LM, Pate MB, Bergles AE (1989a) A comparison of 150 and 300 SUS oil effects on refrigerant evaporation and condensation in a smooth tube and a micro-fin tube. ASHRAE Trans 95:387–397

Schlager LM, Pate MB, Bergles AE (1990a) Evaporation and condensation heat transfer and pressure drop in horizontal, 12.7-mm microfin tubes with refrigerant 22. J Heat Transf 112 (4):1041–1047

Schlager LM, Pate MB, Bergles AE (1990b) Performance predictions of refrigerant-oil mixtures in smooth and internally finned tubes-Part I: literature review ASHRAE. Atlanta Trans 96 (Part1):160–169

Shao W, Zhang Y (2011) Thermally-induced oscillatory flow and heat transfer in an oscillating heat pipe. J Enhanc Heat Transf 18(3)

Shinohara Y, Tobe M (1985) Development of an improved thermofin tube. Hitachi Cable Rev 4:47–50

Smit FJ, Meyer JP (2002) R-22 and zeotropic R-22/R-142b mixture condensation in microfin, high-fin, and twisted tape insert tubes. J Heat Transf 124(5):912–921

Sur B, Azer NZ (1991) An analytical pressure drop prediction model for condensation inside longitudinally finned tubes. ASHRAE Trans 97(2):54–61

Tang L, Ohadi MM, Johnson AT (2000a) Flow condensation in smooth and micro-fin tubes with HCFC-22, HFC-134a and HFC-410A refrigerants. Part I: experimental results. J Enhanc Heat Transf 7(5):289–310

Tang L, Ohadi MM, Johnson AT (2000b) Flow condensation in smooth and micro-fin tubes with HCFC-22, HFC-134a and HFC-410 refrigerants. Part II: design equations. J Enhac Heat Transf 7(5):311–326

Tang L, Ohadi MM, Johnson AT (2017) Experimental results of flow condensation in smooth and micro-fin tubes with HCFC-22, HFC-134a and HFC-410A refrigerants. J Enhanc Heat Transf 24(1–6)

Thomas DG (1967) Enhancement of film condensation heat transfer rates on vertical tubes by vertical wires. Ind Eng Chem Fund 6(1):97–103

Traviss DP, Rohsenow WM, Baron AB (1973) Forced convection condensation in tubes: a heat transfer correlation for condenser design. ASHRAE Trans 79(Pt. 1):157–165

Traviss DP, Rohsenow WM (1971) The influence of return bends on the downstream pressure drop and condensation heat transfer in tubes. ASHRAE Trans 79(1):129–137

Tsuchida T, Yasuda K, Hori M, Otani T (1993) Internal heat transfer characteristics and workability of narrow thermo-fin tubes. Hitachi Cable Rev 12:97–100

Uehara H, Sumitomo H (1985) Condenser. US Patent 4,492,268 assigned to Iisaka Works, Ltd.

Venkatesh KS (1984) Augmentation of condensation heat transfer of R-11 by internally finned tubes. M.S. thesis, Department of Mechanical Engineering, Kansas State University, Manhatlan

Vrable DA, Yang WJ, Clark JA (1974) Condensation of refrigeration-12 inside horizontal tubes with internal axial fins. In: 5th international heat transfer conference, vol 3, pp 250–254

Wang W (1987) The enhancement of condensation heat transfer for stratified flow in a horizontal tube with inserted coil. Int Heat Transf Sci Technol:805–811

Wang HS, Honda H (2003) Condensation of refrigerants in horizontal microfin tubes: comparison of prediction methods for heat transfer. Int J Refrig 26(4):452–460

Wang WW, Radcliff TD, Christensen RN (2002) A condensation heat transfer correlation for millimeter-scale tubing with flow regime transition. Exp Thermal Fluid Sci 26(5):473–485

Webb RL (1999) Prediction of condensation and evaporation in micro-fin and micro-channel tubes. In: Heat transfer enhancement of heat exchangers. Springer, Dordrecht, pp 529–550

Webb RL, Ermis K (2001) Effect of hydraulic diameter on condensation of R-134a in flat, extruded aluminium tubes. J Enhanc Heat Transf 8(2):77–90

Webb RL, Kim NY (2005) Principles of enhanced heat transfer. Taylor and Francis, New York

Webb RL, Yang CY (1995) A comparison of R-12 and R-134a condensation inside small extruded aluminium plain and micro-fin tubes. In: Vehicle thermal management systems proceedings, pp 77–86

Webb RL, Eckert ER, Goldstein R (1971) Heat transfer and friction in tubes with repeated-rib roughness. Int J Heat Mass Transf 14(4):601–617

Webb RL, Zhang M, Narayanamurthy R (1998) Condensation heat transfer in small diameter tubes. In: Proceedings of the 11th international heat transfer conference, vol. 6, pp 403–408

Wu X, Wang X, Wang W (2004) Condensation heat transfer and pressure drop of R22 in 5-mm diameter micro-fin tubes. J Enhanc Heat Transf 11(4)

Xin M, Wang Z, Liao Q (1998) Condensation for steam in the horizontal tubes with three-dimensional inner microfins. J Eng Thermophys 19:464–468

Yang CY, Webb RL (1996) Condensation of R-12 in small hydraulic diameter extruded aluminum tubes with and without micro-fins. Int J Heat Mass Transf 39(4):791–800

Yang CY, Webb RL (1997) A predictive model for condensation in small hydraulic diameter tubes having axial micro-fins. J Heat Transf 119(4):776–782

Yang CY, Fan CF, Chang FP (2003) Effect of fin tip radius for film condensation on micro-fin surfaces. J Enhanc Heat Transf 10(2)

Yasuda K, Ohizumi K, Hori M, Kawamata O (1990) Development of condensing thermofin-HEX-C tube. Hitachi Cable Rev:27–30

Yu J, Koyama S (1998) Condensation heat transfer of pure refrigerants in micro-fin tubes. In: Proceedings of international refrigeration conference at Purdue, pp 325–330

Zhang M (1998) A new equivalent Reynolds Number model for vapor shear-controlled condensation inside smooth and micro-fin tubes. Ph.D. thesis, Pennsylvania State University, University Park

Zhang M, Webb RL (2001) Correlation of two-phase friction for refrigerants in small-diameter tubes. Exp Thermal Fluid Sci 25(3–4):131–139

Zivi SM (1964) Estimation of steady-state steam void-fraction by means of the principle of minimum entropy production. J Heat Transf 86(2):247–252

Chapter 6
Conclusions

- In this research monograph, heat transfer enhancement techniques have been discussed exhaustively for both boiling and condensation.
- Experimental data as well as theoretical analysis available in the open literature have been discussed with all its aspects.
- By and large, surface tension forces are the single most responsible force for both boiling and condensation.
- It has been observed that with the passage of years, more and more sophisticated surfaces have been developed as the evolution of manufacturing techniques has permitted the production of such surfaces.

Additional References

Abu-Orabi M (1998) Modeling of heat transfer in dropwise condensation. Int J HeatMass Transf 41 (1):81–87

Ayub ZH (1988) Pool boiling enhancement of a modified GEWA-T surface in water. J Heat Transf 110(1):226–268

Hiroshi H, Shigeru N, Bunken U, Tetsu F (1986) Effect of vapour velocity on film condensation of R-113 on horizontal tubes in a crossflow. Int J Heat Mass Transf 29(3):429–438

Liebenberg L (2000) A review of refrigerant condensation in horizontal micro-fin tubes. Proc ASME 40:155–168

Marto PJ, Rohsenow WM (1966) Effects of surface conditions on nucleate pool boiling of sodium. J Heat Transf 88(2):196–203

Metals YI (1982) YIM heat exchanger tubes: design data for horizontal rope tubes in steam condensers. Technical Memorandum 3, Yorkshire Imperial Metals, Ltd., Leeds, UK

Traviss DP (1973) The influence of return bends on the downstream pressure drop and condensation heat transfer in tubes. ASHRAE Trans 79(1):129–137

Webb RL (1982) Investigation of surface tension and gravity effects in film condensation. In: Proceeding of the 7th international heat transfer conference, vol 5, pp 175–180

Webb RL (1992) An analytical model for nucleate boiling on enhanced surfaces. Proc Pool and External Flow Boiling Conf 1:345–360

Webb RL, Kim NH (2004) Principles enhanced heat transfer. Garland Science, New York

Index

Printed in the United States
By Bookmasters